ARAGON
ISSUES IN
PHILOSOPHY

D0253454

# PARAGON ISSUES IN PHILOSOPHY

PROBLEMS IN PERSONAL IDENTITY James Baillie

INDIVIDUAL, COMMUNITY, AND JUSTICE: ESSENTIAL READINGS IN SOCIAL AND POLITICAL PHILOSOPHY Jeffrey Barker and Evan McKenzie, Editors

PHILOSOPHY AND FEMINIST CRITICISM: AN INTRODUCTION Eve Browning Cole

PHILOSOPHY OF SCIENCE James Fetzer

FOUNDATIONS OF PHILOSOPHY OF SCIENCE: RECENT DEVELOPMENTS James Fetzer, Editor

PHILOSOPHY AND COGNITIVE SCIENCE James Fetzer

FOUNDATIONS OF COGNITIVE SCIENCE: THE ESSENTIAL READINGS Jay L. Garfield, Editor

MEANING AND TRUTH: ESSENTIAL READINGS IN MODERN SEMANTICS Jay L. Garfield and Murray Kiteley, Editors

LIVING THE GOOD LIFE: AN INTRODUCTION TO MORAL PHILOSOPHY Gordon Graham

PHILOSOPHY OF SPORT Drew Hyland

PHILOSOPHY OF TECHNOLOGY: AN INTRODUCTION Don Ihde

CRITICAL THEORY AND PHILOSOPHY David Ingram

CRITICAL THEORY: THE ESSENTIAL READINGS David Ingram and Julia Simon-Ingram, Editors

SOCIAL AND POLITICAL PHILOSOPHY William McBride

RACE, REASON, AND ORDER: PHILOSOPHY AND THE MODERN QUEST FOR SOCIAL ORDER Lucius Outlaw

METAPHYSICS: A CONTEMPORARY INTRODUCTION John Post

AFRICAN PHILOSOPHY: THE ESSENTIAL READINGS Tsenay Serequeberhan, Editor

INTRODUCTION TO MODERN SEMANTICS Ken Taylor

WOMAN AND THE HISTORY OF PHILOSOPHY Nancy Tuana

**FORTHCOMING TITLES**

HEART AND MIND: ESSAYS IN FEMINIST PHILOSOPHY Eve Browning Cole and Nancy Tuana, Editors

INTRODUCTION TO THE PHILOSOPHY OF LAW Vincent Wellman

# THE PARAGON ISSUES IN PHILOSOPHY SERIES

At colleges and universities, interest in the traditional areas of philosophy remains strong. Many new currents flow within them, too, but some of these—the rise of cognitive science, for example, or feminist philosophy—went largely unnoticed in undergraduate philosophy courses until the end of the 1980s. The Paragon Issues in Philosophy Series responds to both perennial and newly influential concerns by bringing together a team of able philosophers to address the fundamental issues in philosophy today and to outline the state of contemporary discussion about them.

More than twenty volumes are scheduled; they are organized into three major categories. The first covers the standard topics—metaphysics, theory of knowledge, ethics, and political philosophy—stressing innovative developments in those disciplines. The second focuses on more specialized but still vital concerns in the philosophies of science, religion, history, sport, and other areas. The third category explores new work that relates philosophy and fields such as feminist criticism, medicine, economics, technology, and literature.

The level of writing is aimed at undergraduate students who have little previous experience studying philosophy. The books provide brief but accurate introductions that appraise the state of the art in their fields and show how the history of thought about their topics developed. Each volume is complete in itself but also complements others in the series.

Traumatic change characterizes these last years of the twentieth century: all of it involves philosophical issues. The editorial staff at

Paragon House has worked with us to develop this series. We hope it will encourage the understanding needed in our times, which are as complicated and problematic as they are promising.

John K. Roth
Claremont McKenna College

Frederick Sontag
Pomona College

# PHILOSOPHY OF TECHNOLOGY: AN INTRODUCTION

**ALSO BY DON IHDE**

Author

*Instrumental Realism: The Interface Between Philosophy of Science and Philosophy of Technology*

*Technology and the Lifeworld: From Garden to Earth*

*Consequences of Phenomenology*

*Existential Technics*

*Technics and Praxis: A Philosophy of Technology*

*Experimental Phenomenology*

*Listening and Voice: A Phenomenology of Sound*

*Sense and Significance*

*Hermeneutic Phenomenology: The Philosophy of Paul Ricoeur*

Editor

Paul Ricoeur, *The Conflict of Interpretations*

Co-editor

*Descriptions* (with Hugh Silverman)

*Hermeneutics and Deconstruction* (with Hugh Silverman)

*Interdisciplinary Phenomenology* (with Richard Zaner)

*Dialogues in Phenomenology* (with Richard Zaner)

*Phenomenology and Existentialism* (with Richard Zaner)

# CONTENTS

PREFACE

# ON WAYS TO USE THIS BOOK

In developing this introduction to the philosophy of technology, I have tried to follow the guidelines and demands of the Paragon Issues in Philosophy series. Those called for some introduction to the primary literature, problems, and development of the field. And, insofar as any "philosophy of..." has both philosophy and subject matter dimensions, I have written this text in such a way that several strategies may be used.

The most obvious one for any book, of course, is to read it from beginning to end. Were one to do this he or she would find that there is a kind of running narrative which constitutes a general argument about why so much of philosophical history has neglected technology and why this neglect should be corrected. But there are two other strategies which might be equally useful.

One is what might be called a "teacher's strategy." In North America most philosophy courses are introductions and few curricula include sequential or accumulative courses. Thus, there is always a certain, perhaps dominant, number of beginners in every class. Many teachers of philosophy would like to at least expose beginning students to the thinking of our ancestors and philosophical giants. That is part of the reason why I have woven the first chapter particularly with references to some of those thinkers. I am giving an excuse to the instructor to supplement the narrative with bits of the "classics." Plato, Marx, Bacon, Descartes, Aristotle, Kant, Hegel, are all alluded to. And one might well follow the guidelines suggested with relevant readings from those authors.

The other dimension is another narrative about the history of technology. Here distinctive authors are harder to allude to, particularly since the ancient engineers and architects are in many cases not known, nor did they write very much about what they did. But, associated with these "technology blooms," are other philosophical authors who are seldom read in standard classes—yet many of them are very interesting. Archimedes, Aristarchus, Stratos, and Francis Bacon, all might just turn out to be interesting to the instructor as well as to the student.

Indeed, depending upon the class context, it may be just as easy to begin with chapter 2, Technology, as with chapter 1, the Introduction, with its large list of philosophers. For many, the concrete illustrations and examples may prove to be more interesting to explore than texts alone.

Chapters 3, 4, and 5 bring us into the contemporary world. They outline problems which can be discussed and debated. And chapter 4 introduces a suite of authors who are very important to philosophy of technology discussions. In one of my own large classes I divided up the entire class into three groups, assigning one whole book to students within each group, for a report and group discussion in contrast and in arguments with the other two similar groups. The result was a stimulating debate and discussion.

Such a short introduction can only do so much—it is intended as a kind of net thrown out to cover the territory, but it also includes guiding threads within its interwoven cloth, which are meant to suggest different areas of possible deeper exploration and development.

By way of acknowledgments, I should like to thank my own university, SUNY Stony Brook, for providing me with a research semester, and the University of Sydney, Australia, for hosting me during that time. I particularly want to thank the students who read and criticized the early versions of the manuscript while I was "down under." Much of the multicultural background material which is referred to here and in recent works arises out of non-Euro-American travels and experiences.

While I worry about the fate of technoculture—which was so

dramatically made present in this last year during which the Gulf War was fought, the aftermath of which I witnessed as I flew from Europe over Tehran and then over the smoke pall from Kuwait, into the brighter blue skies of the Southern Hemisphere—I also appreciate the delight which can come from new technologies as I see my wife, Linda, learn to like and play with her computer, and my son, Mark, with all the technologies he so quickly masters.

One thing is clear: technology can no longer be taken for granted. It must be addressed. And that is presumably one reason why the philosophy of technology enters this series as a new field of philosophical inquiry.

# PHILOSOPHY OF TECHNOLOGY: AN INTRODUCTION

CHAPTER ONE

# INTRODUCTION

The philosophy of technology is one of the most recent of the sub-specializations in philosophy. Yet, it also belongs to one of the oldest of the disciplines, philosophy, while it is necessarily addressed to a very complex phenomenon, technology. As with any introduction, special decisions are called for in dealing with any subject matter, but particularly in dealing with a new field.

I have chosen to follow a "narrative" or historical story approach to this new field so that both a context for the philosophy of technology may be set, and yet also give a developmental sense to its inception. I shall begin with a tale which draws upon a dominant or "textbook" story which will be familiar to many, regarding first the placement of the philosophy of technology in the *history* of philosophy. That story usually places philosophy from its very beginnings very close to the origins of *science*. But science is not always thought of as having necessary or even close relations to *technology*.

Then, at first implicitly, but then more explicitly, I must also allude to the *history of technology*. And it will appear that not only has the philosophy/science connection not always been interpreted to entail technology, but that the histories of philosophy/science and those of technologies are often quite non-parallel.

One of the enigmas to be addressed revolves around why the philosophy of technology arrives so very late in the history of Western, and particularly North American philosophy. Behind this enigma also lies a very long-standing and deep prejudice which links philosophy and science in a *theoretical* moment and preference. That is part of the set of assumptions which guide the standard or dominant narratives about philosophy and science. But this is not all the story.

I therefore, in telling the tale, also interweave in the dominant themes occluded or forgotten moments which do relate to a different history—a history of technology. This is the implicit story strand which will be interwoven from the beginning. At first it will not be so obvious how and why the history of philosophy and science relates to a history of technology, for the high moments in both do not always coincide, and some tellers of the tale will even accentuate the non-coincidence.

Once the initial historical summaries are told, I shall turn to a more positive interpretation of the role and relationship of philosophy and *technology,* in this opening to a contemporary *philosophy of technology.*

## 1.1 PHILOSOPHY

Philosophy is arguably the *oldest* continuous discipline in the Western canon of academic studies. The very term, academic, is derived from the "Groves of Academe" which was the location of Plato's philosophic "university" which was founded in 385 B.C. But even this ancient Academy had been preceded by a century and a half of "pre-Socratic" philosophers who had begun to develop the critical and reflective style of questioning and reasoning which would become the now twenty-five plus century line of continuation which was to characterize Western *philosophy.*

This is not to say that philosophy in this long tradition always took the same shape. Quite contrarily, even its beginnings were marked by distinct transformations in what it was interested and how it would go about its critical questioning. The standard interpretation—derived from Aristotle's early history of philosophy and cemented in nineteenth century interpretations—has it that the pre-Socratics began philosophy with *nature speculations.*

These early philosophers, usually beginning with Thales (585 B.C.) and running through Parmenides and Heraclitus (roughly 500

B.C.), and up to the origins of ancient "atomism" in Democritus (roughly 400 B.C., a contemporary of Plato), began to make what we today might identify as the first *secular* and *scientific* queries about the nature of the universe. Much later those speculations were named "metaphysics"—again by Aristotle—or that which follows the "physics" which was an ancient term for the powers of nature. But, if this interpretation is followed, it is also easy to see that what would pass for both philosophy and *science* were then inseparable. One could argue that earliest philosophy was "scientific" insofar as the questions were directed towards the "nature" of Nature.

The pre-Socratics, according to Aristotle, were what we would call speculative theorists who argued that *physis* (Nature) had some form of foundation from which all other forms derive—a kind of primeval "water" (Thales), or something "indefinite" (Anaximander), or a kind of "air" (Anaximenes), or, finally, imperceptible, invisible "atoms" (Democritus). Pre-Socratic philosophy is usually identified with those ancient speculations about Nature and here philosophy/science were identical.

That identification did not always pertain throughout the subsequent centuries, but it was frequently revived and most strongly in our long history during the centuries which saw the making of what is today called Modern Science. Modern Science, however, does not mean twenty-first or even twentieth century science—rather it means the historical roots of the Modern Era which for philosophy and science in turn means the Renaissance and following.

Galileo (1564–1642) is frequently taken to be a primary symbolic figure in this history, although he is *late* in the Renaissance. So, if we take 1600 as a time of establishment of Modern Science, then it is the late sixteenth and seventeenth centuries following which are the watershed. From that time on until no more recently than the nineteenth century, the revived interest in nature speculations begun by the ancient Greeks still remained identical with philosophy, although with a clearly identified branch of it, called *natural philosophy*. However, the closeness of philosophy to science can clearly be seen in both the ancient and the renewed, Renaissance and later Enlightenment periods of Western history.

Science, however, is not the only close relative to ancient philosophy. That is already implicit in the terming of the pre-Academeme era as pre-Socratic. Socrates, Plato's mouthpiece, was what could be called a humanist philosopher. He professed the futility of nature speculations and favored, instead, a species of questioning which focused upon *self-knowledge,* "to know thyself" was *first* philosophy for him.

In part this was because pre-Socratic science had reached a kind of apogee in which it could go no farther. Democritus, today recognized as the inventor of the *atom,* at least in concept, could and did speculate upon these presumably imperceptible particles which made up all of Nature, but he could only speculate. Science was as yet not 'experimental' in its modern sense. To be experimental in the modern sense entails a number of factors, including setting up a situation in which certain variables can be controlled; in which a measurement occurs thus implying a mathematical or quantitative judgment about something; but above all and particularly for purposes of this book, *experiment entails technologies or instruments against which and in relation to which the phenomenon is compared.* This is not to say that the Pre-Socratic and Classical Greek/pre-modern scientific thinkers entirely lacked a sense of experiment. But it should be said that most speculation, as brilliant as it often was, was not either in a context of measurement, verification, nor instrumental manipulation. Indeed, if one takes note of the most anticipatory developments of ancient science insofar as they appear to be like modern attainments, *it is not until nearly a century later than Classical Greek philosophy, in the post-Aristotelean period in Alexandria, in which one begins to see the very first anticipations of "technoscience."* I wish here to distinguish more sharply than usual between precisely the Classical Greek and the "Hellenic" or Helleno-Roman developments of early science. I shall return to this period shortly.

Socrates, at the height of the Classical Greek period, and not having at hand any solid accomplishments of the experimental sort hinted at above, could see the folly or the frustration of what we might call the "purely theoretical" or speculative approach to such unexperienced things as atoms, since there was no way to get at them either directly or indirectly. But if this was a negative turn with respect to early philosophy/science, Socrates's humanism also had its positive side: the concern for the highest values possible for human and *rational* knowledge.

Socrates, of course, speaks through Plato (and a few of the other lesser known contempories who knew him). The Dialogues opened the way to another ancient theme, the Good, the True, and the Beautiful. These could be called, retrospectively, ancient *humanities* which were part of the Classical Period of ancient Greek thought. But, like the ancient nature speculations, Socratic (and Platonic) humanism was basically secular and critical. One of the charges brought against Socrates at his trial was that he was "impious" and taught disrespect for the gods. While the Dialogues of Plato are not "atheistic" in that there was at least a finite kind of god, they clearly were not orthodox with respect to the pantheon of the older religio-literary gods of the time.

If both Greek "scientific" and "humanistic" philosophy was critical and essentially secular, philosophy did not stay within either of its early Greek molds. Indeed, just as nature speculations came into being with the Greeks, they also passed out of much of our earlier histories of thought—more so than the Platonistic strands of thought which were soon to take new shapes in the next eras.

As Greece declined and broke up into warring city-states or was invaded by others, philosophy moved into new cultural contexts, Hellenic and Roman. The post-Classical periods, prior to the rise of Christendom, are among the most neglected in our standard histories of philosophy. Yet, in the Hellenic-Roman periods several important changes were also taking place which only in the present may be seen to be important to the subject matter to which we shall shortly turn.

The Hellenic period of our ancient history was a time of great *technological and experimental* development and comes the closest in ancient times to anticipating anything like modern science. Interestingly, its center no longer lay in Greece itself, but in Alexandria which at the time contained the largest library and museum of both the past and what today might be called the multicultural present. It was this group of post-Aristoteleans, approximately a century after Aristotle, who developed this first experimental or 'technoscience' approach to various matters.

This was the time of the legendary Archimedes (287–212 B.C.) who is said to have claimed to be able to move the earth itself were he given a long enough lever. He is claimed to have been an engineering genius who invented a ship-shaker placed under the sea in Syracuse Harbor, as well as other devices. It is doubtful, however, that all of these were actual inventions. They do reveal that the use of technologies was beginning to be thought of in relation to scientific activity. The first treatise on "mechanics," usually attributed anachronistically and wrongly to Aristotle, was most likely written by the Hellenic thinker, Strato (d. 269 B.C.), and is the first treatise on engineering. Material repositories of world knowledge were founded, the most important of which was the famous library at Alexandria. This was the time when many of the "wonders" of the ancient world were built, including the famous lighthouse at Pharos. Much more important for the future, however, were the Hellenic inventions of measurement devices — particularly for time and distance — such as the sundial, *clepsydra* (water clock), gnomon, etc. Nor should one ignore a still later scientific giant, Ptolemy (90–168 A.D.), who also lived in Egypt, the son of Greek immigrants. His was to be the cosmology, the mechanics, the measurements which were to dominate the understanding of the known universe for centuries to come.

For the most part, however, with the possible exceptions of the invention of cement and Roman road building which are frequently cited, most of the engineering feats and other attainments of these centuries has lain forgotten and unattended to. Yet, in the late Hellenic and early Roman periods, many of the features of "modern" living which we take for granted came into being, including glass windows, indoor plumbing and bathing facilities, central heating and cooling, etc. Most, of course, were destroyed in the eras to come, but records, revivals and curiosity over what had been built were to return to the West, although much later and by indirect routes. It might even be suspected that the forgetfulness attending this high Hellenic period and the tendency to anachronistically project it back to the Greeks may have something to do with the persistence of the prejudice for our chosen ancestors and their values in interpreting our own situation today.

Still following the clues of popular history—I am doing this for the sake of common landmarks, even though that process has the danger of taking too much for granted in an oversimplified way—we again know that after the decay of Greece and the rise of first the Hellenic, then the Roman eras, that "Rome, too, fell." The standard histories then term the time the "Dark Ages" in the sense that much of the learning of the ancient and even Hellenic and Roman worlds was lost, destroyed, or—in another of our cultural historical blind spots —was taken into the then emerging *Islamic world.* It is arguable with some cogency to claim that while the northern "barbarians" did all they could to destroy Roman culture, the Mediterranean destroyers of the Empire often preserved its ancient knowledge. Rome itself was dismantled, stone by stone, leaving only its fragments, in the "stone trade" of the Dark Ages. Here, the already cut stones and facings could be recycled into everything from farmhouses to city halls, and that is what happened. Many of the enchanting stone buildings far out in the Italian countryside include stones, marble facings, and materials pirated from imperial Rome itself.

The knowledge, the texts, and the practice of the Hellenic and Roman worlds were often lost, destroyed, or transformed, but not in every direction. This other development in the history of Western technoscience came from an unexpected source: Islam. Muhammad (570–632) began this religio-cultural movement in the very height of the European dark period. What remained of Greek and Hellenic thought was often preserved by Islamic scholars. By the ninth century there was a flourishing development of Islamic science, including the development of an instrumentally embodied science. It was through the Arabs that Ptolomy was reintroduced, and, of particular importance, it was through Arab geography that many of the instruments which allowed for long distance navigation were introduced to Europe.

Indeed, it was to be several centuries later, largely through the Islamic scholars, that most of what we today know of ancient Greek thought, particularly in its scientific mold, reentered Western consciousness.

Meanwhile, philosophy after its Latin and Roman periods, was to undergo another of its historical transformations. It could be said to have turned "theological." Theology—although not religion as such—could be called the hybrid invention of Judeo-Christian religion and ancient philosophy. From the early neo-Platonists, through St. Augustine, into the high Middle Ages with the famous names of Anselm, Aquinas, and Abelard (associated with the revived universities which arose from 950 to 1250 in Europe) philosophy was to concentrate upon a religio-theological subject matter. In the twentieth century this transformation has been called "ontotheology." And, just as ancient philosophy could in some sense be thought of as identical with science, so, in the high Christian eras, philosophy was virtually identical with religious thought in the form of theology. Not that this identification was always a comfortable one: many of the great European debates from the fourth through the sixteenth centuries centered around the interpretations of "faith" and "reason," read "religion" for the first, "philosophy" for the second terms of the binary. And, just as had happened in the Socratic compared to the pre-Socratic eras, by the late Middle Ages into the emergence of both Protestantism and the pre-Renaissance, there was another humanist revolt. These humanists—Roger Bacon (1214–1292) who wrote of the famous machines da Vinci was much later to draw, Desiderius Erasmus (1466–1536) at the birth of the Renaissance, and others—were to set the stage on one side for what was to become a revival of ancient concerns which became, in turn, the birth of Modern Science. The humanists encouraged a revival of basically secular knowledge and of critical reasoning related to letters, rhetoric, and historical and literary criticism. They were also instrumental in getting print into the vernacular, and hence no longer solely in the hands of an elite which performed only in the dead language Latin.

The other pre-Modern source for Modern science, was again the occluded Other of so much European history, the Islamic scholars. It was through Islamic philsophers such as Avicenna (980–1037) and Averroes (1126–1198), even earlier than our humanists and more closely interested in science than our ontotheologians, who knew, commented upon, and preserved the ancient Greek texts and knowledge. And in the two-century "pluralist" period in Spain, Moors, Jews, and Christians collaborated in intellectual ventures which gradually leaked the scientific concerns of the ancients back into European thought.

This knowledge and these texts included Aristotle, Ptolemy, and the physician, Galen. But it was to be Aristotle who came to dominate European philosophical thought for several centuries among the ontotheologists. A pupil of Plato, in the ancient context, Aristotle could be thought of as the great systematist and classifier of all knowledge. He, unlike Socrates and Plato, found it easy to combine *both* nature speculations and humanist concerns. And while Plato had remained at least partly continuous in our intellectual history as belonging to the Neo-platonic and later ontotheologies of Christian thought, Aristotle had to be revived. It was in the pre-Renaissance revival of Aristotle, and later associated with him some of the Greek materialists, that interests in science were to revive and give birth to Modern Science.

Aristotle's fate, however, was to be ironic. It could be said that the nature or science-oriented thought which through Aristotle was re-introduced in the pre-Renaissance, was to itself be inverted and become a protagonist for much of early Modern Science. In short, while the early inventors of Modern Science were passionately interested in a science of nature, their direction later was one which often took as its theme, *disagreement with Aristotle.* This occurred with Galileo and the interpretation of physical motion, and later with Descartes across an even broader spectrum. But however the tables revolved, in the Renaissance and following, philosophy again could find its identity close to and in a sense encompassing "natural" science.

During this time, beginning with the revivals of travel and building which followed the reestablishment of a European version of the Empire, another occluded history was occurring—the Middle Ages were another *technological revolution.* This was happening concurrent with the rise of universities, from the tenth century on, in which philosophy was debated in theological terms. The cities, the cathedrals, the draining of the Lowlands of northern Europe were made possible by ever larger and larger *machines* such as cranes, wind and water mills, and devices using simple physical principles, some of which had been recovered from Roman and Hellenic, but also Asian, sources. Indeed, by the time of the Renaissance, the "mechanization" of Europe had already occurred and sophisticated feats of engineering such as the Duomo in Florence could be completed. It was not an accident that such "artists" as Leonardo da Vinci and Michelangelo were as much "engineers" as artists—and in ways to be noted later—which also could be taken for granted by a father of Modern Science, Galileo. It is not yet apparent what this occluded history has to do with philosophy in its scientific, revived form. But philosophy, too, had changed.

Yet, not all was ever to be the same or simply that. Through its various transformations, philosophy had become too comprehensive and diverse to be simply identified with either science, humanities, or theology—in a sense, it was and was not all of these. Its critical and rational style of thinking could be said to have been related to each of these divisions of concern. By the post-Renaissance, "natural philosophy" was clearly recognized as a distinct and distinguishable *area* of philosophy, or a "speciality" within philosophy. But neither were the sciences, theology, or the humanities yet genuinely separate from philosophy in the more contemporary sense.

We now may move to another era in the common historical narrative I have been following. This common interpretation has it that as the various sciences matured, they did begin to become distinct, *and to break away from philosophy per se.* Such emergence *from* philosophy was seen, at first, as positive by both philosophers and scientists. By the eighteenth, but much more in the nineteenth century, philosophy began to elevate its concerns to higher or more "transcendental" ones. This was to be the time of the great *metaphysical* systems. The names of Immanuel Kant (1724–1804) and G.W.F. Hegel (1770–1831) stand out here.

Science—now we might call it primarily *physics*—had progressed in the fields, particularly, of astronomy and physics in the modern sense as the study of motion and matter. Isaac Newton (1642–1727) was the great systematizer of much that had gone on, and became the successor to Galileo. In his time, Newton had called himself and was called a "natural philosopher," but with Immanuel Kant, such terminology was to change. The system of Kant's *Critique of Pure Reason* was to become a *metaphysics* of science, an interpretation and critique of what had been accomplished by Newton as a "scientist." Philosophy had begun to differentiate itself *from* science. If science was *physics;* philosophy was *metaphysics.* As *meta*physics, Kantian philosophy saw its role as a critical interpretation of "what makes science possible" or giving the "grounds of the possibility" of science.

Here was Aristotle's metaphysics as "after the physics" clearly elevated to metaphysics as "above physics." But I shall use a less exhalted metaphor: In the *Critique of Pure Reason,* Kant's metaphysics of Newtonian science, philosophy became a lot more like literary criticism to a novel, or art criticism to a work of art, than it had been as a *producer* of literary works or artistic ones. For example, Gottfried Leibniz (1646–1716), not long before Kant, was simultaneously a scientific creator—he was one of the inventors of the infinitesimal calculus—and a philosopher. But with Kant, the task of philosophy in this critical role became that of interpreter, critic, and appreciator of some field or other.

Here was the latent beginning of the next transformation, the invention of *"philosophies of..."*. Hegel accelerated this process by becoming both more distinct from given philosophical subject matters, and more elevated into a *meta*-realm. Hegel was not satisfied with the already grand spectrum of critiques of "pure" and "practical" and other forms of reason as was Kant, but elevated his metaphysics to the whole of history itself. Hegel's was the grandest of the systems and set the tone for virtually the whole of the nineteenth century. The stage is now close to being set for the contemporary, the twentieth century.

## 1.2 PHILOSOPHIES OF...

Hegel quite explicitly began do do "philosophy of..." (in German, of course, the preposition "of" does not occur. But *Geschites-philosophie* or *Religionsphilosophie,* both terms used by Hegel, when translated into English become "philosophy *of* history," and "philosophy *of* religion"). Following the above characterization, a "philosophy of..." looks at a subject matter and thematically, critically interprets it and analyzes it.

Historically, several other features of nineteenth century philosophy should also be noted. First, by that time philosophy as a distinct discipline was recognized—there were chairs of philosophy as distinct from other disciplinary chairs. Similarly, in science, there were also recognized disciplines—although far fewer than would be the case now in the twentieth century. But philosophy, still following its ancient and comprehensive aims, could and did take on critiques or analyses of any discipline whatever, thus also there could be a proliferation of "philosophies of..."

Interestingly, in Europe this included at least one foray into *Tecknikphilosophie,* a book by the neo-Hegelian, Ernst Kapp, in 1877! *Yet, it was to be nearly 100 years later that philosophy of technology became a recognizable characterization within philosophy in North America.* As late as 1979, a well reputed philosopher of science, Mario Bunge, could remark in his own developing interest in philosophy of technology:

> Technophilosophy [Bunge's term for philosophy of technology] is still immature and uncertain of its very object, and does not exploit the entire

> scope of its own possibilities. That it is an underdeveloped branch of scholarship is suggested by the fact that so far no major philosopher has made it his central concern or written an important monograph on it.[1]

As we shall soon see, this claim is overstated and, in context even untrue—but Bunge was not without reason in making such a claim. *Philosophy of technology*—particularly when compared to such much older "philosophies of..." as religion, history, or even science—is clearly a relative *newcomer to the North American scene.* Why should this be the case? To see why we must return to the story of philosophy's history, now as it enters the twentieth century.

## 1.3 THE TWENTIETH CENTURY

Here we return to the narrative concerning philosophy and science once again. If the nineteenth century was the time when philosophy was separated from and distinguished from science, the now "parallel" history of disciplines began to show something else—science had attained a kind of success and ascendancy which it had not had heretofore. And contrarily, philosophy, particularly in the proliferation of "systems" of metaphysics, had reached a kind of impasse. Both neo-Kantians and neo-Hegelians dominated the universities and, while very busy developing vast systems of metaphysical interpretation, had retreated into a kind of grandiose academic self-enclosure.

The sciences, on the other hand, were beginning to be of service to both a whole range of "applied" disciplines, and were implicated in the development of the Industrial Revolution. Science's investigation of the phenomena of magnetism and its technologies, led to the electrical technologies of the twentieth century. The newly emergent science of chemistry—arising out of the technologies of dying processes in Germany—began to be one of the models for the industrial/corporate Big Science of the late twentieth century. And by mid-century, the now established nomenclature which relates the "pure" or "theoretical" sciences such as physics, to the "applied" sciences such as engineering were already in place.

This apparent success of the children of philosophy—the sciences—compared to the apparent dead end of proliferating and conflicting metaphysical systems brought about what many early twentieth-century philosophers called a *crisis* for philosophy. This was the case with the three styles of philosophy which were destined to become competitors for Euro-Anglo-American philosophy by the mid-twentieth century—*Pragmatism, Positivism, and Phenomenology.* These philosophies set the style for twentieth-century philosophizing much as had Kant and Hegel for the eighteenth and nineteenth.

The figures or groups associated with these movements were John Dewey, pragmatism; Edmund Husserl, phenomenology; and the "Vienna Circle," sometimes including the enigmatic Ludwig Wittgenstein, with Positivism. All three movements shared a deep suspicion of metaphysics, at least in its tendency to construct apriori systems of everything. And all three movements shared the recognition that science and scientific methods had risen to the fore of rational and critical thinking, as well as exhibiting a success in the explosion of knowledge heretofore unknown in human history.

These two features translated into a much more *modest* form of philosophizing for the early to mid-twentieth century. Philosophy was to become more "problem oriented," more "concrete," and more "analytical." But it was to retain its interest in "philosophies of..." Of course, each of these three contenders had its own mode of *doing philosophy,* a catch phrase used by each of these movements.

Pragmatism strived to be problem engaged. Dewey's corpus included treatises on art, education, and science-technology, as well as a host of other topics—he became first known as America's foremost philosopher of education, but as we shall see later, he was also important to what has become contemporary philosophy of technology. For him, philosophy itself would be interpreted as an activity of intelligence of a "pragmatic" or "instrumental" nature.

Positivism took a more guarded approach, ceding to science itself virtually the whole realm of "empirical truth," thereby restricting philosophy to a new mode of *logical and linguistic analysis,* but also, having made science the center of the epistemological quest, Positivism placed a distinctive cast on twentieth-century *philosophy of science* in and through the mid-century. Philosophy would be distinct from science, but would be a kind of handmaiden clarifying the processes of reasoning within science.

Phenomenology took as its task the remaking of philosophy as a new "rigorous science of experience," and developed what we can now recognize as a method of relativistic and ecologically descriptive analysis. And while Husserl eventually was to play an important role in late-twentieth century philosophy of science, it was to be one of his younger colleagues, Martin Heidegger, following a phenomenological model, who was to be a primary developer of philosophy of technology.

What is important to note at this point, however, is that the spectrum of these three styles of philosophizing again transformed the role and self-understanding of philosophy. Descending from the "heights" of metaphysics, philosophy in the form of the "three P's" became engaged in problem-centered, particular problems.

Not to be forgotten or occluded here, is the dominant factor which also arose in the twentieth century—the rise of maximal and high technology. If early Modern Science was preceded by the beginnings of the world voyages and the discovery of the New World and the rise of science as distinct disciplines accompanied by the machine age of the Industrial Revolution, the twentieth century saw science become fully a *technoscience* now thought to be the "motor" which drove and developed what is called Modern Technology, presumably distinct from any ancient or traditional technologies.

## 1.4 PHILOSOPHY OF SCIENCE

We shall now take what may, at first, seem something of a detour. If philosophy of technology is a very new arrival in North America, the *philosophy of science* is one of the very earliest of the "philosophies of..." to become organized and institutionalized. The

Philosophy of Science Association was formally organized in 1934 and today is one of the largest of the sub-disciplinary groups within philosophy on this continent.

Given the story of the history of philosophy and science just traced, that is perhaps not surprising since the history of philosophy and science seems so interwoven. Yet, philosophy has also changed its shape over its long history, as we have seen. By the 1930s it is not philosophy overall which is linked with science overall, but a philosophy *of* science which relates to a subject matter—science. The scene we have found at the early part of the twentieth century is one in which philosophy within the "three P's" has reacted to the grand metaphysical philosophies of the nineteenth century and now wants to investigate its highly successful offspring, the sciences. But while there are representatives of all three of the "P's" involved at the birth of the North American Philosophy of Science Association, soon it would be dominated by one of the "P's," Logical Positivism. And, more, it would not be all the sciences which stand at the center of this interest, but primarily the natural sciences, particularly *physics, and that in its purest "theoretical" form.*

There was a short as well as a long historical reason why this should be the case. The short, or immediate reason lay in the fact that physics was seen as the most mature and successful of the early twentieth-century sciences—and it was soon to become the first truly Big Science operation during and immediately after World War II with the Manhattan Project.

Associated with this perceived success was also the theory of science itself adopted by Positivism. This philosophical school of thought held that at the heart of knowledge lay a kind of pure reason exemplified by certain modifications of *logic, and logical reasoning.* Early philosophers of science proclaimed that science, unlike all other ancient forms of knowledge, was genuinely progressive and grew in truth overall. Its means was that of hypothetical-deductive reasoning which Positivists saw most highly exemplified in physics. Indeed, this preference for a logical and hypothetical-deductive model of what science is, sometimes led the more extreme proponents of Positivism to exclude many branches of what commonsensically would be thought to be science, from the canons of science itself. Geology was one such example, but certainly also those sciences which bordered on the social sciences, such as primatology, or behavioral psychology, or even paleontology or physical anthropology, were too speculative and not purely deductive enough to count as genuine sciences.

All this belonged to the *early* history of Positivist philosophy of science, and while such prejudices continue to exist in some attitudinal corners of the universities, Logical Positivism as the dominant strand of philosophy of science was itself to eventually recede into history—but it did dominate and set the tone for much of what was to follow until the mid-twentieth century and what was to be a revolution in the philosophy of science in the 1960s.

Logical Positivism was both an extreme form of philosophy of science, and of a certain tradition concerning how philosophy continued to be closely linked to science. Its focus was that of a *theory-bias*. That bias points to a much deeper and longer set of *prejudices* which lay behind the late arrival of our subject matter, the philosophy of technology. Let us now look at some deeper reasons for the first arrival of philosophy of science, compared to the late arrival of philosophy of technology.

The deeper reasons to be explored may also be termed three contemporary *prejudices* which are closely related to the dominant strands of the history I have followed to this point:

1) The first prejudice is the most ancient and deepest—it is the prejudice which links both philosophy and science to *theory* in an exaggerated way. This prejudice goes all the way back to the Classical Period of philosophy itself taken in a certain way. Were we to try to characterize what the Greeks called *theoria* in a contemporary way, we might say that it combined a kind of imaginative speculation with an attitude of respectful contemplation.

We have already noted that from the pre-Socratics through Plato this use of speculation, first directed toward Nature, then toward Ideas, began to separate early philosophy from its literary, religious, and mythological roots in ancient Greek culture. But, at the heart of *theoria* in the Greek sense was also a kind of quasi-religious belief which associated the "highest" virtues with a contemplative attitude. Existentially, the aim of highest humanity was, in effect, to achieve the deepest knowledge and, once attained, meditate upon it contemplatively. Such was a "pure" thinking which was itself philosophy. This formulation was shared in different ways among the Greeks, but is perhaps exemplified in its strongest form by Plato and becomes, in the long history cited, its "Platonistic" side.

One should have noted by now that the phenomenon *technology* is virtually missing from discussion. But it also is missing from much of the discussion of both philosophy and of science. The "science" of our latest arrivals, the Positivists, is virtually a technology-less science. It is here that we must see how the theory exaggeration is part of this occlusion.

2) The second prejudice is much more recent and contemporary. It associates with the often forgotten "history" of technology in the West. It is a prejudice which holds that Modern Technology is both essentially different from all ancient or traditional technologies, and therefore in some fundamental sense, "better." Indeed, some go so far as to indicate that only those complicated and "hi-tech" machines of the present are really *technology as such.*

3) And the third prejudice is closely interwoven with the second. It is the belief that Modern Technology has, as one of its major differences from all other technologies, been *largely derived from Modern Science.* The institutionalization of engineering the contemporary sense as "applied science" in many universities is an instantiation of this belief.

Were it not for the previous references to occluded histories, the dominant story told so far would seem to lead to and confirm these three beliefs. Let us briefly return to Plato to get a glimpse of how this story might unfold:

Plato has now been used as an exemplar of certain views. And while his *Republic* might seem to be very far from the philosophy of technology, it can be seen to be a kind of model for the thinking which downgrades the material and elevates a kind of contemplative "theory" to highest status. In the *Republic* he gives an account of knowledge called "The Divided Line" which is illustrated in his famous allegory of the cave.

The cave tale is an allegory of enlightenment in which humans are to be found viewing what we might today take as a primitive version of a cinema theatre. They are chained in an audience position, looking up at the wall of the cave at (shadow) images which have been cast upon it by figures being moved along a parapet behind which is a fire (the "projector"). The plot of the allegory is one which revolves around the viewer gradually becoming aware of the causes of what he or she sees. At each stage the discovery of what is causing the images to seem the way they appear takes the viewer ultimately out into the sunlight, outside the theatre and thus into the full light of (truthful) day.

The Divided Line interpretation is that these images are the poorest sort of knowledge, mere copies of the real things. But perception does not do much better, and only when the enlightened ones reach the formal or abstract or theoretical clarity of ideas—above and outside embodiment—can genuine knowledge begin to be reached. It is from Plato that the long and deep Western distinction between appearance and reality takes much of its characteristic shape.

For Plato there was a hierarchy of knowledge, in which the highest object of knowledge was the Good itself, as an abstract and ideal form known only to pure mind through intelligence *(noesis)*. Less abstract was the world of mathematical objects known through thinking *(dianoia)*. But neither the Good nor number were concrete and only below the line drawn under the ideal would that which was visible occur. This world of perception—only perceived—could be believed *(pistis)* as an opinion. And lower still was the world of copies and images which were matters of imaging *(eikasia)*.

In the allegory of the cave which illustrates this hierarchical structure of knowledge, enlightenment begins when the captive cinema-goers are gradually released and begin a journey which first shows them the mechanism for casting shadows, then a trip eventually out to the sunlight which, at first, blinds them (these are the philosophers).

For our purposes it is important to note the values associated with the levels of knowledge, with images and perceptions or, *anything associated with the body and the material,* as inferior to the leap above that line into the realm of *"ideas" or "forms" which presumably associate with mind or soul.* Disembodied knowledge is taken to be superior to embodied knowledge. And this is the contemplative knowledge of the Good, the True, and the Beautiful, but as a slightly lower form, it also includes the mathematical. These values dominated much of Western epistemological history. And the same downgrading of crafts could also be associated with this tradition.

Insofar as both the dualism of body/mind and the valuation of material/immaterial pervades Platonism it also colored much of the history of philosophy/science itself. When philosophy of science took shape, it did so with this set of beliefs and values and until very recently tended to hold that technology—in its contemporary and high-technology forms—derives simply from theoretically and formally constructed science as a theory activity.

As we shall see, philosophy of technology, both later and differently parented, would arise from another tradition—that in which *praxis* takes a higher value and more important role.

All of these beliefs could be said to be shared in some degree by many of the philosophers up to the twentieth century we have cited. And for that reason, the linkage of philosophy and science might be termed theory-biased with respect to its selfinterpretations. This makes the arrival of the philosophy of science in North America seem quite natural. It is part of what remains of the earlier closer identity between a theory-prone philosophy with a theory-interpreted science.

Where, in all of this, is *technology*?

## 1.5 PHILOSOPHY AND TECHNOLOGY

By now it might seem as if philosophy has had little to say about or do with technology. If philosophy is almost solely preoccupied by *theory,* and its history of high theoretical moments disjunctive with those of technology, this might seem even more to be the case. But here my deliberate following of a sort of textbook history must begin to be qualified.

In that textbook history, the high moments of theory-prone philosophy as related to science might have seemed to be (a) Classical Greece, (b) the late Renaissance period of the rise of Modern Science and the rise of Modern Philosophy, and (c) the elevation of great metaphysical systems in high post-Enlightenment Europe which saw in science its ultimate success. Contrarily, low moments in such theory-prone philosophy as related to science, might coincide with precisely those which in our occluded technological and engineering history occurred.

The Classical Greeks were *not* strong in technological or engineering feats. They did accomplish a number of inventions, often associated with warfare or the theatre, such as reputed solar mirror condensers for starting fires on ships, or machines (deus ex machina) for elevating or lowering gods on the stage. Most of their technologies were, in effect, captive to their aesthetics. Thus Pythagorean tuning elevated a five-tone scale which matched the proper proportions of mathematical harmonics according to Greek aesthetic sensibility, and the construction of columns on temples, making them tilt in such a way as to seem upright parallels in perspectival correctness, but in actual construction angled, illustrate this aesthetics-bound engineering. Indeed, as L. Sprague DeCamp points out in *The Ancient Engineers:*

> Like the Egyptians, the Greeks persisted in using the architectural forms that they were accustomed to from the days of wood. All the details of the entablature, with its cornices, friezes, and so forth, were copies in stone of wooden structural elements. Greek builders even imitated the pegs that held the ancestral wooden structure together by adding little stone knobs called 'drops.' [2]

While on the surface this might seem simply a failure to adapt to new materials, this imitation of wood in stone indicates something much deeper in the Classic culture. One should remember that even statuary of finest marble was painted. It was this "illusion" or "copy" imitation practice which so upset Plato in his theory of knowledge. He was, in a sense, responding to a kind of technical praxis in his idealist reaction. The contrast between Classical Greece and its aesthetic and the much later Renaissance can be noted in the dramatic change in style of statuary. It was the material—usually high quality marble—which was glorifed in its naked appearance. Michelangelo could remark about bringing out the shape of that which was latent in the marble, the material.

I have already referred to the higher technology periods following among the Hellenic dispersion of Greek culture and the subsequent Roman period with its far vaster and more innovative engineering discoveries, many of which still dot the European and Mediterranean countrysides. Everything from the Roman arch, which the Greeks did not invent, to aquaducts of spectacular size, to the aforementioned development of modern housing technologies, to fanciful hydraulic systems for fountains, all belonged to this period.

Similarly, while philosophy was little concerned with science in the Middle Ages, even though its logical and theoretical skills were being refined, after the renewal of scholarship and a European "empire," the much more thorough technologization in Europe proceeded. As the historian Lynn White, Jr., has amply shown, the Middle Ages were a virtual technological revolution which borrowed from every conceivable source, but which also innovated in anticipation of the later Industrial Revolution.[3]

And in this later post–"Dark Ages" technological revolution, a feature which also characterized the Hellenic-Roman blooming of technology, could also be found featured in the Middle Ages. That is, cultural borrowing and adaptation played a very strong role. White indicates that the travels taken by traders and crusaders brought back with them new techniques and technologies which were adapted to particular European power needs. Thus, the windmill, borrowed from India and Iran, was harnessed as a mill, a pump, a toolshop. Arches, from the Middle East, took up higher and higher support functions in buildings. Windmills and water-mills, heavy lifting machinery, the architecture of the Gothic with its flying buttresses and pointed arches made stonework virtually frilly in its lightness and height. But, here, too, philosophy preoccupied with theology, seemed to have little to do directly with technology.

In the Renaissance there began to be a much closer relation to technology and certainly it accepted new technologies, although the use remained somewhat implicit. As already noted, the leading precursors of Modern Science were as equally fascinated by technology as by nature. Leonardo da Vinci's incredible (and unworkable) designs of machines for warfare, for flying, and for submarine purposes, are illustrative of this fascination. The great Renaissance artists were also as often engineers as artist-scientists. The same applies to Galileo, who took the use of the compound lenses already extant, and made them into a telescope with which he began to make the great empirical discoveries which belong as deeply to the origins of Modern Science as do the use of ancient mathematics and theory rediscovered from the Greeks via the Moors. His "machines" were also used in his deliberate experiments. He constructed inclined planes with reduced friction tracking. And measuring devices were also developed. Galileo was among the first of the Europeans to make a *technologically embodied science* in his use of instruments and experimental devices for experiment. Galileo *was not a Greek speculator, but a Modern prototype for technoscience.*

Still later, when philosophy was beginning to become grandiose, as in the metaphysical period, Western technology also became grandiose in the Big Machine era of the Industrial Revolution. It was during this period that early Modern Science begin to achieve both the successes noted, and to find itself much more deeply related to technology. In fact, some of the most astonishing discoveries of science—such as the Laws of Thermodynamics—began to be discovered and worked out, not with respect to "bare Nature" but with respect to puzzles arising out of machinery—technology. In this case, thermodynamics arose over puzzles posed by the *steam engine.*

Today, of course, it is inconceivable to think of science without its technologies, ranging from its embodiment in instrumentation to its gigastructure in corporate-modeled Big Science. Here is a wedding of science/technology, which has become, as Bruno Latour terms it, *technoscience*.[4]

Have philosophers failed to notice this? The answer is a qualified no, but even their notice must be placed in our story's context, because until the twentieth century the phenomenon of technology remained a *background phenomenon*.

Plato and Aristotle clearly were appreciative of what they called *technē*. But can *technē* be translated as technology? *Technē* was simultaneously a craft and art object—it could be a marvelous shield, a finely wrought statue, or a vessel for drinking. Art and technology were not separated, and, indeed, intrinsic to the judgment of any such object was not simply its utility, but also its beauty. But, *technē* clearly was a produced artifact, a virtual "technology."

What the Classical Greeks thought, however, falls within the aforementioned "aesthetically determined" technology. If the artifact was produced with purpose, care, and above all the proportionality of Greek sensibility's values, then it was excellent as *technē*. I have suggested that this notion of a narrow technology may be precisely one of the reasons why Greek technology never matured or proliferated. Moreover, no matter how excellent the product, craft/arts skills must take second or third place to the purely contemplative or ideal related *theoria* of the philosopher.

In contrast, one of the reasons technology could begin to proliferate and develop in the later Hellenic and Roman Periods, may precisely be seen in the breakdown of the values exemplified in the narrower ideology of the Greeks. The Hellenic-Roman times were philosophically *eclectic*, reflecting both the high degree of cross-cultural climate of the times, and the relative tolerance which went with it.

Hellenic times were multicultural times. The Alexandrians, as noted previously, were often immigrant Greeks living in Egypt. These were also the culturally eclectic times surrounding the large political and religious transformations which gave rise to Christianity, diaspora Judaism, and the multitude of Oriental and

Roman cults. Trade, ideas, goods, but above all, practices and techniques were interchanged and with the interchange came the technology bloom of that period.

Similarly, in the Middle Ages, much of what was developed consisted of borrowings, modified for European power purposes, from other cultures. Nor should one forget that the often occluded Others—particularly the Arabic and Far Eastern cultures—were playing very strong background roles at this very time. Not only trade, but wars are part of the crosscultural exchange which relates to technological changes. This was also the time when one might say that engineering fantasies began to take place, which were to set up another side of our technological past.

As early as the mid-thirteenth century, Roger Bacon, one of our aforementioned humanists, claimed:

> It's possible to build vessels for navigating without oarsmen so that very big river and maritime boats can travel guided by a single helmsman much more swiftly than they would if they were full of men. It's also possible to construct machines for flight built so that a man in the middle of one can maneoever it using some kind of device that makes the specially built wings beat the air the way birds do when they fly.[5]

And, although Bacon produced no drawings to show these imagined inventions, that is what Leonardo da Vinci was to do—and he had read Bacon—with his more concrete imagination in the sixteenth century! What da Vinci provided was the engineer's visual imagination and representative skills to do the drawings which had often earlier remained merely conceptual.

Roger Bacon's later namesake, Francis Bacon (1561–1626), a younger English contemporary of Galileo, also began to envision what was to become technoscience. Bacon began to transform the *contemplative* ideal of Classical science into the instrumental and interventionist science of the Modern period. He envisioned an interventionist and *experimental* science, a science which necessarily entailed technologies.

Bacon has been credited with the association of knowledge with power: "knowledge is power." Here begins a distinct movement away from the pure theory and contemplative ideals of the Greeks into the modern notion of power knowledge in which the purposes of knowledge are not so much to simply know nature, but to change things. What may not have been obvious in the Baconian change, however, is that power must be applied in order to have knowledge—that was what was implicit in the experimental and technological notion of early Modern Science.

Bacon had become dissatisfied with both the contemplative ideals of the Greeks and the dominant scholastic arguments of the times and opted for an empirical, but interventionist science. Nature had to be forced to yield her secrets: "The secrets of nature reveal themselves more readily under the vexation of art than when they go their own way."[6] And such knowledge was to be considered not only true, but useful: "Truth, therefore, and utility are here the very same things, and works themselves are of greater value as pledges of truth than as contributing to the comforts of life."[7] And, for our purposes, it is important to note that Bacon is among the first to explicitly note that experimental science is necessarily a technological or instrument-mediated science. Beyond the use of telescopes and inclined planes (Galileo), Bacon lists some twenty-seven kinds of experiments and the instrumental devices which go with them, including what he called "evoking devices" which aid perception but also get at previously unperceived phenomena, "reduce the non-sensible to the sensible: that is, make manifest, things not directly perceptible, by means of others which are."[8] Here we begin to have the makings of a science/technology linkage which only much later will become a major theme in philosophy of technology.

Early Modern Science thus is doubly linked to technology—first in the close association of engineering and instrumentation found to be so natural in the Renaissance, and, again, in the transformation of the ideals of knowledge into experimental, interventionist forms of power/knowledge in its Baconian sense which is, again, instrument-embodied. And, here, philosophy begins to be self-conscious about technology. But Bacon was fated, historically, to become a background figure in early modern histories of philosophy, taking a minor role when compared to the theoretically more elegant thinkers such as René Descartes (1596–1650) whose "geometrical method" and elegant arguments of a purely deductive form took the foreground role. Descartes tended to dislike actual technologies and instruments because they failed to deliver the precision of deduction. Like the Greeks, he worried about imitation. It was Descartes who introduced the long-historied worries over whether or not we could be fooled by cleverly conceived automatons, which are, after all, only motioned versions of Plato's same worries over painted statuary. The dominant strands of philosophy usually tend to read Descartes over Bacon, but for that very reason much of the instrumental and technological side of actual scientific practice is overlooked.

In our story, after the rise of Modern Science—which followed the technologization of Europe in the high Middle Ages—comes the age of the large machines, the Industrial Revolution. Its effects were to spawn a different sort of philosophical reflection, but one which did directly relate philosophy *to* technologies.

I have already referred to the first use of a "philosophy of technology" in Ernst Kapp's use of *Technikphilosopie* in 1877. His neo-Hegelian response to technology was largely one of admiration, seeing in technologies another way of confirming the Hegelian insight that to find genuine self-knowledge, Mind *(Geist)* must dialectically relate to its other and become aware of itself within an objective world. But it was to be another neo-Hegelian, who would see in this other a different effect, Karl Marx (1818–1883). For our purposes, this crucial nineteenth century philosopher should be recognized for contributing three major ideas relating philosophy to technology. First, Marx can be thought of as one of the primary inventors of a *praxis* philosophy, that is, a type of philosophy which reevaluates theory and relates theory to more basic levels of action and materiality.

Marx is often said to have "stood Hegel on his head," which means that whereas Hegel had seen Mind or Spirit as a progression in human and social self-awareness arising out of a historical dialectic of ideas, Marx—writing in the milieu of the Industrial Revolution—argued that ideas, ideology, always relates to a set of concrete, material conditions; it is out of practices that ideas are formed. The shapes of human actions—particularly socio-political ones—are what make philosophy possible.

The second idea which Marx introduced, and the one which relates philosophy quite directly to technology, is that among the various material forces which need to be reflected upon are the *material modes of production* which are, in turn, shaped by *technologies.* Here the phenomenon of technology begins to enter philosophical consideration in a positive way.

Marx's various stages of the modes of production are at least superficially well-known. He recognized that for vast stretches of human history, such modes of production changed little or slowly (as with the Stone Ages when tools continued to be made according to long entrenched and stable practices), but at times there were revolutions or changes in the modes of production among which the best known were listed by Marx as (a) the feudal, (b) capitalist, and (c) the to-be communist phases.

Modes of production were, in Marx's sense, systems of material products produced by technologies arranged in different ways. These were embedded in different socio-political structures. The feudal system was one in which there was a set of mutual relations between the lords and the serfs such that the production of food, clothing, and the various goods needed were to be the responsibility of the serfs, while protection, governance, and even possible expansion of a territory were the responsibilities of the lords.

But, insofar as production could and did exceed basic needs—producing surplus value—other modes of life could also begin to occur. The guilds and apprenticeship structures which began to occur in late feudal times, and which led to a middle class and cities, were eventually to erode the feudal structure and give rise to a money economy and early capitalism. By the nineteenth century—Marx's time—capitalism had become industrial, with modes of production much more powerful and with much higher outputs than anything possible under feudalism.

In spite of this growing surplus, there was, thirdly, also a negative underside relating to modes of production—*alienation.* For example, in the early guild apprenticeship system, a producer could be closely related to the entire process of production and the fruits of his or her labor returning to the maker the value produced. But in a factory system of production, workers could be alienated from both their satisfaction in making a full product, but the value produced returned only in part in the form of wages, thus doubly alienating the maker from both the satisfaction of making the product and from its material reward. Such a phenomenon was clearly related to the concrete processes of production and their various shapes.

This *praxis*-centered, *mode-of-production* analysis of society was to be one of the primary sources for one side of what would become philosophy *of* technology. Technology had become, in Marx, not a background, but a foreground issue. And while Marx's more utopian hopes—for a non-contradictory and nonalienating form of productive society, communism—did not take shape according to his hopes, his tradition of analysis did continue. In the twentieth century the Critical Theorists or the so-called Frankfurt School, whose best known representative is Jürgen Habermas, would also become relevant to philosophy of technology.

However, Marx's very breadth, which saw technology embedded in a wider social and political praxis, also leaves technology as *phenomenon* still a dimension of primarily socio-political issues. Furthermore, the Marxian concerns veer away from the theory/science issues which we have located within the textbook history just traced.

We have now again reached the twentieth century. But between Marx and the early to mid-twentieth century philosophers more explicitly interested in technology, stood the two World Wars and the communist revolutions. All three of these events had been changed by the introduction of Industrial Revolution–honed technologies. Classical warfare which centered upon a professional military engaged in close combat in isolated battle areas reached the stalemate of the trenches in World War I. Poison gas, aerial warfare, the submarine, and high-powered explosives began to replace the lo-tech swords-turned-rifles of up-close warfare even then. World War II began to introduce genuine higher-tech processes and histories of engineering still rate that period as the highest period of *technological innovation* in history. In terms of recent engineering history, World War II produced many of the inventions and developments which were to drive the Euro-American technological engine for several decades.

The full mobilization of nations to produce radar (UK and US) against mass bombing attacks (the "Battle of Britain" which was the air battle defending England against the first massive use of air bombing by Hitler's airforce in the 1940s), total city destruction by fire bombing (Dresden, an Allied retaliation in which an entire city was destroyed), and, eventually, the Manhattan Project to produce the atomic bomb, subsequently unleashed upon Hiroshima and Nagasaki, followed on the heels of the social changes of the Industrial Revolution, making technology as a force too important to overlook. European philosophers, particularly, began to make technology and technological civilization a primary theme of their reflections.

These included important contributions by an expatriate Russian, Nicolas Berdyayev, Spain's best known philosopher, José Ortega y Gasset, and the German, Martin Heidegger. These early philosophers of technology wrote immediately before, during, and after World War II. In America, it was to be John Dewey who also made technology a forefront phenomenon. Among these early philosophers interested in technology, it must be said that much of the European reaction to technology was *negative*.

Berdyayev and Ortega saw in modern technology threats to high culture, liberal education, community, and humanity in some of its traditional, classical sense. They argued that there was a kind of leveling result, a technocratic result built into the historical development of technology such that only a species of "mass man" could thrive. Dewey, on the American scene, was to give a more positive evaluation to technology, which he did *not* differentiate from science. Indeed, Dewey saw science as a mode of technological thinking.

After World War II the negative evaluation of technology would, among some European philosophers, become even more stringent. This was particularly the case with two influential mid-twentieth century thinkers: Jacques Ellul (France) and Herbert Marcuse (Germany, then United States) and their respective books, *The Technological Society* (1963) and *One Dimensional Man* (1964). A third thinker, Jürgen Habermas, took a less totally negative tack with respect to technology, although he continued to follow a neo-Marxian trajectory. His *Towards a Rational Society* (1970 [originally 1968 in German]) is, in part, a response to a kinship line.

Both Marcuse and Ellul took technologies into what could be called Technology with a capital letter—they reified all technical products and the processes associated with them into *Technology.*

Technology was taken by both Ellul and Marcuse as a term which associated with calculative and analytic thinking, a kind of twentieth century technological style which threatened to subsume all other styles. Technology, also taken to have become autonomous and no longer under human control, aimed towards a *totalization* of its form.

In Ellul's case, the rise of technique was a mode of thinking which supplanted all pre-modern forms of thought, but, more, it aimed to *substitute* its realm for that of nature: "Technique has become the new and specific milieu in which man is required to exist, one which has supplanted the old milieu, viz. that of nature."[9] In short, technological culture has succeeded in absorbing nature. Ellul himself outlines his program in a series of the characteristics which he claims for the technological order:

a. It is artificial;
b. It is autonomous with respect to values, ideas and the state;
c. It is self-determining in a closed circle. Like nature, it is a closed organization which permits it to be self-determinative independently of all human intervention;
d. It grows according to a process which is causal but not directed to ends;
e. It is formed by the accumulation of means which have established primacy over ends;
f. All its parts are mutually implicated to such a degree that it is impossible to separate them or to settle any technical problems in isolation.[10]

This was really to have revived a very much older notion which contrasted culture and nature, or the "artificial" with the "natural." Reified, such a technological gestalt was seen to be indefeasible, autonomous, and self-determining. It is an extreme form of technological determinism. Technology with the "T" thus replaces its sub-dimensions such as economics, politics, and even culture.

Ellul asks whether the technological society can be civilized and gives a basically negative answer. First, technological society is thoroughly materialistic: "The technical world is the world of material things; it is put together out of material things and with respect to them. When Technique displays any interest in man, it does so by converting him into a material object."[11] What does grow under this society is effective power: "Technical growth leads to a growth of power in the sense of technical means incomparably more effective than anything ever before invented, power which has as its object only power in the widest sense of the word."[12] And, finally, while Technique frees humans from many of the old order's restrictions (material), Ellul contends that the result is the opposite of freedom: "Technique can never engender freedom."[13]

Marcuse, too, saw in Technology the near victory of analytic or technical thinking. Humans in the context of a technological civilization, being reduced to a mono-dimension, or one-dimensional being. He traced this movement with respect to everything from consumer society with its standard products and the transformation of everything into commodities, to the reduction of choice to that between brands, and even to philosophy itself which at the time of *One Dimensional Man* was dominated by the late Positivist versions of analytic philosophy.

But with Marcuse the insight was primarily one which saw in the rise of Technology a kind of technocratic ideology. In the guise of neutrality, what technoscience hides, is a form of dominating power, political power.

The genius of the technological society's new form of control, which Marcuse contends is a new form of totalitarianism, is to have found a way of controlling society without repression.

For 'totalitarian' is not only a terroristic political coordination of society, but also a non-terroristic economic-technical coordination which operates through the manipulation of needs by vested interests. It thus precludes the emergence of an effective opposition against the whole.[14]

The older age of repressive totalitarianism is replaced by the age of total administration.

To be sure, all basic needs (food, shelter, health) are satisfied in the modern technological society and needs now become non-necessary needs. But all is controlled within a cooptive system of needs manipulation and a system of "deceptive liberties as free competition at administered prices, a free press which censors itself, free choice between brands and gadgets."[15]

With Marcuse it is not so much the nature/artificial distinction which animates his characterization of one-dimensional technological life, but of a closure of the political possibilities in the totally rationalized society:

The main trends are familiar: concentration of the national economy on the needs of the big corporations, with the government as a stimulating, supporting, and sometimes even controlling force; hitching this economy to a world-wide system of military alliances, monetary arrangements, technical assistance and development schemes; gradual assimilation of blue-collar and white-collar population[s], of leadership types in business and labor, of leisure activities and aspirations in different social classes; fostering of a pre-established harmony between scholarship and the national purpose; invasion of the private household by the togetherness of public opinion; opening of the bedroom to the media of mass communication.[16]

The technological society thus gains its totality through a certain kind of apparent satisfaction of needs, manipulated through technical means.

Both Ellul and Marcuse belong in different ways to mid-century forms of technologically deterministic interpretations of technology. Bearing some lineage to Marx, they saw not a communist state, but a negative utopia or dystopia emerging from technological development. In this respect, Habermas as a related but critical respondent to the same traditions, sees a different possibility.

Habermas's *Toward a Rational Society* (1970) contained a detailed response to and critique of Marcuse (and by extension, of Ellul). He took the situation to be considerably more complex and less advanced towards closure than either Marcuse or Ellul. Although agreeing with Ellul and Marcuse that the forms of power and technology had changed from pre-Modern times, and that the interpretation of technology could no longer be merely instrumental but taken as a system, Habermas held that there remained a number of features which strongly separated what he called the "social life-world" and the realm of technical control.

The former remains open, communicative, and operates informally, while the technical world has features not easily made practicable in the life-world sense. He contrasted these two sets of features in the following ways. The realm of symbolic interaction remains open. It is characterized by social norms, intersubjectively-shared, ordinary language, reciprocal expectations about behavior related to social roles, and maintains itself institutionally. Emancipation remains an internal goal of such interaction. In contrast, technical or instrumental and strategic sub-purposive groups follow technical rules within a context-free special language and set up conditional predictions and imperatives. Action is defined in terms of special skills, an orientation to problem solving, and the maximization of efficiency and sheer power of technical control. But such groups remain limited to special areas within a larger socius.

What Habermas did note in depth was that technology in modern times, particularly after the rise of the industrial state (primarily in the nineteenth and twentieth centuries) became both capitalist, or corporate, in form and bound in new ways to science. First, with respect to society overall:

> In this way traditional structures [of society] are increasingly subordinated to conditions of instrumental or strategic rationality: the organization of labor and of trade, the network of transportation, information, and

communication, the institutions of private law, and, starting with financial administration, the state bureaucracy.[17]

And, in a more momentous departure from most neo-Marxians, Habermas notes that:

Since the end of the nineteenth century the other developmental tendency characteristic of advanced capitalism has become the increasingly momentous: the scientization of technology.... Thus technology and science become a leading productive force, rendering inoperative the conditions for Marx's labor theory of value.[18]

And, with what today must appear to have a certain predictive power, Habermas concluded that:

In consequence of the two tendencies...capitalist society has changed to the point where two key categories of Marxian theory, namely class struggle and ideology, can no longer be employed as they stand.[19]

What emerges in this societal form of late capitalist technology is what Habermas called "technocratic consciousness": "The leading productive force—controlled scientific-technical progress itself—has now become the basis of legitimation."[20] Thus, at the very heart of late modern society, technologically embodied science has been installed.

With this tracing of our story, once again, into the middle of this century, we have at least seen philosophers begin to take the phenomenon of technology seriously. But this is still philosophy *and* technology, it is not yet what I am calling philosophy *of* technology.

Indeed, in this reading, we have reached a possibly dangerous point precisely because what we have seen in our textbook history has been a close, and largely *positive* relationship between philosophy and science, but now we have begun to introduce technology more explicitly into the discussion and it seems to appear to have a *negative* relation.

We have also seen that the twentieth century subspecialization which takes science as its foreground phenomenon, the philosophy of science, has paid little attention to technology per se, although there have been repeated recognitions in its history that at least so far as science is concerned, at least its instrumentation is highly important. Francis Bacon's recognition of this was, if anything, elevated in the early twentieth century by Alfred North Whitehead, who saw much of the key to the success of science due to its technologies, instruments:

> The reason we are on a higher imaginative level is not because we have a finer imagination, but because we have better instruments. In science the most important thing that has happened in the last forty years is the advance in instrumental design...[21]

The danger is that if philosophy has historically been closely and *positively* associated with science and continues to be so in the philosophy of science, then if its relations to technology were to be seen in the light of the early twentieth-century philosophers,' *negative* reactions, we would clearly exacerbate precisely the prejudices of the textbook history we have followed. But now it is time to turn in this introductory chapter, finally, to the *philosophy of technology.*

## 1.6 PHILOSOPHY OF TECHNOLOGY

To qualify as a philosophy of technology in the sense in which I am using it in this introduction, the philosopher must make technology a foreground phenomenon and be able to reflectively analyze it in such a way as to illuminate features of the *phenomenon of technology itself.*

We have seen the emergence of a philosophical recognition of technology, both in the modern tradition at the rise of science, and in the nineteenth- and twentieth-century reactions to growing technologization of society. But for a clearer beginning of the philosophy *of* technology, I turn to the early part of the twentieth century and two different, but related seminal thinkers: Martin Heidegger (Germany, 1889–1976) and John Dewey (United States, 1859–1952).

Both were what I have already called *praxis* philosophers, that is, thinkers who found a unique kind of knowledge associated with action or patterned practices. And both were to relate this kind of knowledge to technology and *a technological way of doing or seeing,* thereby formulating types of philosophies of technology. I begin with the European of our pair, Heidegger.

There is something of an irony in this narrative in that three "idealists"—Plato, Hegel, and now, Edmund Husserl—are formulators of philosophies which stimulated inversions of a more 'materialist' sort which could lay the basis for the philosophical concern with technologies. Aristotle, reacting to Plato's exclusion of the sciences, became more scientific; Marx, reacting to Hegel's radical claims for the precedence of Mind, produced a dialectical materialism, and, as we shall see, Heidegger, in response to Husserl's analysis of "pure consciousness" produced a powerful analytic of ordinary praxis.

Husserl's phenomenology, a name echoing Hegel's earlier version, began as an attempt at describing a realm of pure ideas. And, like Hegel, it was dialectical in the sense that the kind of knowledge which could be arrived at was only via an interaction with the world. But this interaction was actually a kind of philosophical relativity: all knowledge is reflexively relative to an actual and *embodied* experiencer (or observer). Thus in a move similar to Einstein's in which the position of the observation must always be taken account of, Husserl made the concreteness of position central to his mode of knowing. Such an epistemology *enhanced both perception and bodily position* and thus inverted much Platonism which denied to perception and position any primacy.

Martin Heidegger, Husserl's younger colleague, saw the implications of this shift and applied it to a concrete, existential realm which would give birth to the first serious "philosophy of technology" in the contemporary context. Already in his first major book, *Being and Time* (1927) he argued that scientific knowledge—contrary to the dominant positions of the time—was itself a *derivation from* and *dependent upon* a more primitive *praxical* knowledge.

Drawing from Husserl's insistence that all knowledge was constituted from a concrete, bodily position and a perceptual base, Heidegger argued that our primary relations to a world or experience environment were not first conceptual, but praxical, bodily relations which are exemplified in ordinary activity.

In an example which was to become one of the most cited passages in twentieth-century philosophy, Heidegger used the illustration of a hammer to make his argument. A cobbler (such as his father) picks up a hammer *within a relational context of specific uses* in which the hammer "belongs" to a kind of taskfield. It implicitly refers to its place in this context—referring to nails, to the soles of the shoes, to the production of the artifact. But the hammer, in use, is not an "object-as-such." And its actual use displays a peculiar kind of practical knowledge which is not primarily conceptual, but bodily.

One must know how to use a hammer, but once having learned, the hammer in use withdraws as an object and becomes the means of the experience itself. Were we to change the hammer example to a more contemporary artifact, we might use the word processor to illustrate the same point.

When typing, our attention is upon the display terminal. We watch the appearing product taking shape and *do not explicitly* attend to the typing itself. Our intended actions flow through the machine virtually without notice and the machine, in order to be well functioning, must allow this withdrawal. Only when the machine malfunctions, or is broken or missing, does it obtrude as an object impeding our intentional action and thus becomes an ob-ject—that which stands against. It is this *disruption* to the flow of praxis that Heidegger claims provides the occasion for a different kind of knowledge, knowledge of objects as other than or over against our insertion into the praxical life we ordinarily experience. And such a non-use knowledge of objects can become itself a theme of what potentially might become science. What is important to note, however, in this early insight, is that technology plays a role in this primal human experience of an environment, but in such a way that it is both taken-for-granted and in such a way that it may appear to be functionally (if functioning well) virtually invisible.

Several decades later, in an essay, "The Question Concerning

Technology," Heidegger was to draw a far more extensive set of conclusions from these insights. He began to argue in the post-War period, that ontologically, *technology precedes science.* That is, he explicitly *inverted* the values of the dominant traditions regarding theory and practice, and science and technology.

To do this, however, he had to reconceive technology. In the hammer example, it might seem that technology is a tool, albeit a tool which in use becomes transparent and which belongs to a tool-context, the work project. Yet, the work project also implies a kind of world or environment. In his earlier *Being and Time,* Heidegger did hold that ultimately such a practical environment implied a certain kind of world in which everything gets accounted for in some way.

Thus a covered railway platform takes account of the weather (snow and rain) within a human-world context. It was this taking account which Heidegger raised to a holistic model, technologies, which in his later works on technology—became Technology (capitalized to emphasize its global character). Technology with the "T" became a systematic way of seeing the world. That way of seeing a whole world was characterized by Heidegger as a framework, within which the totality of the world could be seen as a kind of resource well (the usual translation of *Bestand* is "standing reserve") in which all of nature becomes an energy resource for human instrumental use. It is this technological way of seeing the world which characterizes the whole Modern Age. In such a way Heidegger may be said to be a progenator of a philosophy which raises Technology to utmost importance in understanding the human world.

Before leaving this early "philosophy of technology," there is one final point to take note of regarding Heidegger's inversion of the science/technology relation, which now becomes the Technology/science relation. Heidegger argues that the very use of mathematics, experimental method, and the scientific abstraction—theory itself—must be seen as a kind of "tool" of the Technological way of seeing:

> It has been said that modern technology is something comparably different from all earlier technologies because it is based upon modern physics as an exact science. Meanwhile we have come to understand more clearly that the reverse holds true as well: modern physics, as experimental, is dependent upon technical apparatus and upon progress in the building of apparatus.... [But] Modern physics is not experimental physics because it applies apparatus to the questioning of nature. The reverse is true. Because physics, indeed already as pure theory, sets nature up to exhibit itself as a coherence of forces calculable in advance, it orders its experiments precisely for the purpose of asking whether and who nature reports itself when set up this way.... Because the essence of modern technology lies in enframing, modern technology *must* employ exact physical science.[22]

Here we have a full inversion of the Platonic view and the birth of a distinctive philosophy of technology. This birth, however, was not Heidegger's alone. Across the Atlantic his American counterpart, John Dewey, was also reformulating philosophy in a technological—or instrumental—direction.

John Dewey was an older contemporary of Heidegger, and given both the thirty years seniority and what clearly is an arguable precedence for notions important to the philosophy of technology, it might seem that he should have been dealt with first. I am not doing so for the simple reason that his role in this history was itself only to be discovered later in the North American uses of both Heidegger and Dewey.[23]

One possible reason for this lateness of discovery of Dewey as a pioneer in the philosophy of technology lays precisely in the closeness of Dewey's thought to a technological way of seeing—or in terms closer to Dewey himself—a technological mode of inquiry. It could be said that Dewey was the first to model *philosophy itself upon a technological model of inquiry.* But Dewey himself was to make this identification only late in his career. In the earlier part of his career, Dewey's philosophy was identified with *pragmatism,* although he preferred the term "instrumentalism" for his mode of analysis. But in 1946 he indicated: "It is probable that I might have avoided a considerable amount of misunderstanding if I had systematically used technology instead of instrumentalism in connection with the view I put forth regarding the distinctive quality of science as knowledge."[24]

Dewey's instrumentalism saw human inquiry, insofar as it focused upon knowledge, as a problem solving, expansive mode of inquiry very like all practical or *praxical* action. Philosophy, science, and *technology* as such a process operated similarly, if not identically.

Like Heidegger and also like Ludwig Wittgenstein, one of the principals in his earlier period of Logical Postivism, Dewey was one of the transformers of twentieth century philosophy, reacting against the apriori and metaphysical system building of the nineteenth century. And, he too, reworked the theory-bias of traditional philosophy, by arguing for a certain primacy of *praxis* in ways similar to those already identified in Marx and Heidegger: "All controlled inquiry and all institution of grounded assertion necessarily contains a *practical* factor; an activity of doing and making which reshapes antecedent existential material."[25]

Looking at the entirety of the history of philosophy, and of science in this way, Dewey was led to lower the distinction between both philosophy and science and the instrumental or a *technological* mode of practice. In a somewhat less dramatic fashion than Heidegger's claim for the ontological priority of technology over science, Dewey claims that science itself is essentially experimental/technological:

> There is no ground whatever upon which a logical line can be drawn between the operations and techniques of experimentation in the natural sciences and the same operations and techniques employed for distinctively practical ends. Nothing so fatal to science can be imagined as elimination of experimentation, and experimentation is a form of doing and making...[26]

> Indeed, it may be said that the distinction between science and *other technologies* is not intrinsic. It is dependent upon cultural conditions that are extrinsic to both science and industry. Were it not for the influence exerted by these conditions, the difference between them would be conventional to the point of being verbal.[27]

And, Dewey even argued that the experimental/technological nature of science itself could become an illusion similar to what we have noted in Heidegger's argument:

> Controlled inference is science, and science is, accordingly, a highly specialized industry. It is such a specialized mode of practice that it does not appear to be a mode of practice at all.[28]

Here, too, then is an inversion of much of the theory-bias of traditional philosophy, such that technology may become a foreground phenomenon for explicit philosophical investigation. The problem, however, in the Deweyan context, was that his very philosophy itself was so closely modeled upon an understanding of technology, that it was only several decades later that he was to be reclaimed for the philosophy of technology.

I have now situated the *late* arrival of philosophy for the philosophy *of* technology. But if Dewey and Heidegger may be seen as two pioneers in this history, the uses of their work to make philosophy of technology an institutional form similar to that occupied by other "philosophies of..." had to wait until a later part of the twentieth century.

Technology as a theme did emerge into prominence by the mid-twentieth century. Its first arrival, however, could be said to have stood under a deep suspicion and worry. Ellul and Marcuse, with their reified and negative views, were widely read in the 1960s. But *in the 1970s and 1980s a virtual explosion of philosophical works on technology began to appear.*

At first there were anthologies and compilations of articles, including an early issue on the possibility of the philosophy of technology as early as an issue of *Technology and Culture* in 1967. But among the first to systematically begin to relate philosophy and technology, were two sympathizers for Dewey, but also for Ellul, Carl Mitcham and Paul Durbin. Mitcham, with Robert Mackey, edited *Philosophy and Technology: Readings in the Philosophical Problems of Technology* (1972). Informally, there began to be sporadic conferences on "philosophy and technology" which would eventually lead to a formal organization only in 1983 when the Society for Philosophy and Technology was organized (there still is no formal "philosophy of technology association" to parallel the PSA).

Much of this early work was related to an interdisciplinary movement within the colleges and universities for curricular developments in Science, Technology, Society or STS courses and programs. Philosophy was finally becoming self-aware about technology.

This rise in interest saw a number of important books published, such as Langdon Winner's *Autonomous Technology* (1977), Edward Ballard's *Man and Technology* (1978), William Barrett's *The Illusion of Technique* (1978) but, as Bunge observed earlier, "philosophy of technology" as a deliberate focus, with that title, began to appear only in 1979. George Budliarello and Dean Doner edited, in 1979, *The History and Philosophy of Technology* and, as a first in a recognized philosophy of science series, my *Technics and Praxis: A Philosophy of Technology* (1979) appeared, shortly followed in the same series by Friedrich Rapp, *An Analytical Philosophy of Technology* (1981). Following this, many more books began to deal with philosophy of technology, including introductions such as *Philosophy of Technology* by Frederick Ferré (1988) and, now, this book (1993). A first North American series in the *Philosophy of Technology,* as contrasted to the older *Research in Philosophy and Technology* series first edited by Paul Durbin and later by Frederick Ferre, published its first books in 1990: Michael Zimmerman, *Heidegger's Confrontation with Modernity,* Larry Hickman's *John Dewey's Pragmatic Technology,* and my *Technology and the Lifeworld.* Nor should Albert Borgmann's *Technology and the Character of Modern Life* (1984) and Winner's second book, *The Whale and the Reactor* (1986), be left out. In short, from the 1970s on, philosophy of technology began to take its place alongside the other "philosophies of..." which characterize much of the contemporary philosophical thrust for critique and analysis.

Before leaving our story, now that technology has been explicitly recognized by the philosophy of technology, simply take note that there are two implicit anomolies which may seem to exist between the "philosophy of science," and the "philosophy of technology." First, the parentage of these two sub-specializations is clearly different. Where we left the philosophy of science in its early North American thrust, it was dominated by Logical Positivism, which later became largely Anglo-American or "Analytical" in style. And it was largely unconcerned about technology per se.

Contrarily, philosophy of technology can be seen to be dominated by adherents of phenomenology (Heidegger) and pragmatism (Dewey), or, from political traditions (including the later Wittgenstein), with others coming from neo-Marxian (Critical Theory) and some from theological backgrounds, but all from versions of "praxis philosophies."

Secondly, whereas the early North American philosophy of science we have spoken of so far has been dominated by its Anglo-American roots, it has also been largely positive both toward theory-bias, and toward science. But our philosophers of technology, while clearly making technology a foreground issue, have often been critical and guarded toward its modern ramifications.

We shall return to this story again.

## CHAPTER TWO

# TECHNOLOGY

Any "philosophy of…" is directed toward some subject matter or phenomenon. Thus in a philosophy of technology the phenomenon to be investigated, reflected upon, and critiqued will be *technology.* Yet, one may not simply stipulate some definition of technology without also setting possible arguments and directions which could either skew or foreclose various avenues of inquiry. Definitions are *not neutral.*

For that reason, we shall begin here with a very broad, but also a concrete preliminary definition of technology. First, we shall insist that a technology must have some concrete component, some material element, to count as a technology. And, second, a technology must enter into some set of praxes—"uses"—which humans may make of these components. And, third, we shall take as part of the definition, a *relation* between the technologies and the humans who use, design, make, or modify the technologies in question.

As broad as such a definition is, it nevertheless is at best middle-sized with respect to contemporary discussions of technology. It is considerably narrower than the definitions which make technology equivalent to any calculative or rational *technique.* For example, in contemporary sports, various techniques have been created to achieve higher performance and, in many cases, *there is a clear technological component.* If running or playing is video-taped, then broken down to frame-by-frame analysis in order to find the most efficient form of motion, it is clear that a technology has been employed to perfect a technique. The technique, in this case, is clearly technologically implicated, although it itself is not a technology. Moreover, many techniques need not employ technologies—styles of speech, modes of courtship, habits of observation, in the past and apart from our more saturated technological feedback, technologies have all been techniques without technologies in this sense. Today, each of the above might well employ technologies to enhance the learning involved and thus parallel the sports case which does implicate, *indirectly,* technologies.

At the same time, this middle-sized definition is not so narrow as to preclude counting as technology any of the historic or even prehistoric technologies which are pre-Modern. It is anthropologically-philosophically broad enough that most forms of *material culture* will be seen to be related to technology. Moreover, in its preliminary stage, it also does not stipulate that a technology needs to be made or manufactured per se.

If one mythically retroprojects into pre-history a "first" technology, it might well have been a *found technology.* A stick picked up and used as a club, or a broken gourd used as a container by a prehistorical man or woman would be a found technology. This is also the kind of technology which many animals use, although I shall call animal uses proto-technological.

This is not to say that such natural objects used as technologies were not, virtually from the beginning, *modified* or designed to fit a given praxis better than a merely found object. Primatologists have discovered that primates such as chimpanzees already incorporate such design features into their prototechnologies. Chimps shape the grass probes that they use for getting termites out of mounds and even make several before use, knowing that the grass probes will quickly wear out. The simplest spears actually used by so-called Stone Age peoples were, in fact, quite elaborately manufactured: the shaft was smoothed and straightened; the point hardened, usually by fire; any added point or spearhead, manufactured from bone or stone, was attached by a vegetable glue and/or cord; etc. The step from ideally shaped found technologies to minimal and then maximal "designing" was a major, but short step.

This anthropological-philosophical inclusion of so-called primitive technologies is quite deliberate and important. We are, of course, long familiar with the notion of *homo faber,* humans, the makers. And while it is clear that *there are no human cultures which are pre-technological*—all humans have a material culture with complexly patterned praxes involving artifacts—we have only recently begun to appreciate the *complexity* of even what may be called technologically minimalist cultures.

We now recognize much more fully than we once did, that pre-historical artifacts such as stone and bone tools, are but a small part of the tool kits of pre-historic peoples. Many of the most important artifacts would have been lost or have deteriorated over the millenia involved. While "Acheulean hand axes" have been discovered from southern Africa to northern Europe in designs which preserve a set process over many, many millenia, it is now not doubted that there were lost tools such as baskets, nets, lines or ropes, containers (gourds, eggshells, etc.) which allowed our forebears, both men and women, to engage in their lifeworld praxes. Asia, for example, no doubt had a multi-millenia bamboo culture stretching far back into pre-history, for which little remains by way of the extremely narrow material artifacts available for archeological analysis.

While all cultures were thus technological, there were also a very high variety of ways in which distinct cultures *embedded* their technologies into lifeworlds. Let us take culinary technologies as a preliminary example. In broadest outline, almost all cultures had a variant of the bowl, and most of a spoon, for dealing with liquids (soups, stews, drinks, etc.), and most had some sort of "stabbing" device for spearing small bits of hot foods. But only some cultures developed the fork—even in our own lineage, the fork was relatively late. Medieval culinary technology had a dagger for stabbing small fowl, but hands and fingers were used to tear these morsels apart.

In contrast, the use of chopsticks, very ancient, developed also only in part of the world. And the cooking gear which went with the styles of preferred cuisines also varied and were differently derived. It is surmised that the wok may have evolved from using a breast-plate as a cooking device for stir-frying on the march. The fork evolved from a two-pronged dagger, better made to keep the morsel from twisting or dropping off the single blade form. Moreover, the preferred style—skewers and barbeques in various cultures, contrasted to stir-fry or steaming in others, or the eternal stewpot for still others—historically exemplified at one and the same time, distinct culinary styles employing different technological complexes.

Once again, the point here is that within the long human-technology history, there is a "universal" occurance of human-artifact relations, but a particular and culturally diverse set of praxes which revolve around the same processes (cooking, storage, preservation, etc.) and the same would be seen to be the case with respect to very divergent activities. Archery, for example, was independently developed at many points of history and in different parts of the globe. And while it may seem that the bow and arrow are the same technology, in fact they are very differently made, used, and even differently contexted.

Certain tribes in jungle areas use a pinch technique to fire the arrow up into a heavily leafed overhead for a monkey or bird target. The bow is usually relatively short and so is its range. In contrast, the ancient longbow of the Anglo-Saxons is fired by a four finger bowstring pull, and can deliver an even armor piercing penetration with rapid fire. The Zen archer, whose whole philosophy of firing is that of the arrow firing itself, uses yet another variant upon archery. And, finally, it should be noted that in Australia, the bow was never invented, but instead an incredible variety of boomerangs—only specialized ones of which return to the thrower—was developed to serve the same purposes. Again, we see both the same technologies differently embedded, or different ones to serve similar purposes, but the praxes which extended human capacities are instantiated in each cultural variant. Technologies will be seen to be deeply *culturally embedded.*

To this point I have used pre-Modern examples, in part to illustrate that human-technology relations have always been part of our human context, but also to show that technologies have had a wide variety of pre-scientific and even pre-industrial development. I shall continue to move slowly into a more concentrated focus upon the contemporary situation precisely because it will also show that many of the problems which we associate with Modern Technologies must also be seen in a broader context. I shall utilize the same technique used in describing philosophy in that I shall re-tell parts of the story of technology in ever more pointed fashion.

## 2.1 TECHNOLOGY AND THE ENVIRONMENT

Why technology in the first place? The answer, anthropologically and philosophically, revolves around humans relating to their *environment,* whether conceived of as a small territory, or more largely, as contemporarily, to the Earth itself. Even without technologies animals, and ourselves as human animals, modify at least local environments. Ant lions modify the very local environment of the sand by making the funnel shaped holes into which their prey may fall just as humans, in building shelters, modify environments, whether minimally as with desert nomads, or maximally as with contemporary urban dwellers.

Technologies, however, allow this modification to be *amplified or magnified.* This *non-neutral, transformative power of humans enhanced by technologies* is an essential feature of the human-technology relations we shall examine. This feature of technologies in human hands is one of the most obvious features of the present and is implicated in the current *environmental crisis.* The amplifying/magnifying power of technologies, in the late twentieth century, has brought to the fore the human-technological power of a geological force. One of the illustrations of such geological force brought about by industrialized technologies currently discussed is the Greenhouse Effect, which has been reported to include changes in the atmosphere of up to 24 percent of certain gases of "homogenic" origin. On a much smaller scale, but less debatable, are the disasters which occur through the uses of mega-technologies such as super tankers and the spills of petroleum and petrochemical products such as that of the Valdez incident in Alaska in 1989, or worse, the eco-terrorism of the Gulf War in which oil spills and oilwell fires were deliberately unleashed in the conflict. Much of the contemporary debate around and about technology revolves around such events.

Yet, it is important from the beginning to place such powers and results in context, for humans, even using the simplest of lo-technologies, have often made major modifications to the environment. Two of the groups of the most monumental ancient engineering developments, the 4500-year-old pyramids of Egypt and the 2500-year-old Great Wall of China (still the only human structure visible from the Moon), were accomplished with very simple machines. Environmentally destructive effects have also often been the result of lo-tech or even no-tech actions of humans. Much of the desertification which is today accelerating in parts of the globe is caused by the use of sheep and goats in sparse foliage areas. These animals graze too closely for grass recovery and their hooves often degrade the soil surface, hastening erosion. Or, to take an example closer to Western civilizational origins, all of the Mediterranean peoples—Phoenicians, Hebrews, Greeks, and Romans—practiced the same deforestation and overgrazing practices which led to the current climate and ecology of the Mediterranean. That basin was, even in historic times, once richly forested with soils amidst the now virtually barren rocky hills, and covered with springs and other water sources, now dried up since the rise of our Greco-Roman past. None of these examples of environmental degradation depended upon modern or hi-tech processes. Still, with the much greater amplificatory/magnificational powers of modern technologies, it can easily be seen that the same processes applied to the extant rain forests could cause their loss *in a much shorter time*—the centuries required for some of the above examples, could, today, be reduced to decades.

It would be tempting, at this point, to immediately plunge into one of the macro-problems surrounding the phenomenon of technology in the late-twentieth century. Is environmental degradation on a global scale a *result* of Modern Technology? Or, is any technology simply neutral and dependent upon simple human use, with good results dependent upon good uses and bad results upon bad uses? And although I shall resist this temptation a bit longer, it should be seen already that the two extreme forms of the debate are oversimplistically stated.

If we stick to our concept of technology as always related to humans, then any human + technology will depend upon more than mere use, but also upon the latent powers of the particular technology involved as related to the complex of possibilities open to the human involved. Secondly, insofar as technologies are culturally embedded, the variant cultural traditions will also come into play. Even at this minimal level it should be seen that every aspect of the human-technology equation is one which displays non-neutral transformational possibilities.

In the range of examples suggested, it can be seen that humans + lo-technologies can after a long enough time effect large environmental territories. Deforestation, irreversible erosion, regional climatic changes, have historically so happened. The same effects can occur rapidly in more technologically maximal and hi-tech contexts. Neither of these results yields a view of technology as neutral or simply a tool of human purpose, but the results do show great differences in how those results can be achieved. Nor do these differences relate only to obviously negative results.

One of the most obvious *positive* results arising out of modern technologies relates to much of what we take for granted with respect to a standard of health which is relatively disease-free. Slow, accumulative results which effect this state relate to the growth of hygienic practices in a great many areas—the isolation or removal of harmful organisms through cleanliness techniques relating to food and water supplies, as an example—compared to the quick and dramatic fixes which occur through the development of medical solutions such as the Salk vaccine for polio, a disease which ravaged tens of thousands of people in the days of today's students' parents or, at most, grandparents. These are positive developments as dramatic in time frame as the threats of only decades to go with respect to rain forests! Both positive and negative effects, however, point to the greater amplifactory and magnifacational powers of modern as compared to ancient technologies.

If all technologies, ancient or modern, minimal or maximal, have implications for some range of environmental territory, the philosophical problem will be to isolate what are the *variable* and the *invariable* features of these transformations. Moreover, the analysis must be one which takes into account, not one, but several dimensions of the phenomenon: (a) the nature of the various technologies involved, (b) the relation or range of relations to the humans who use (and design or modify or even discard) them, and (c) the cultural context into which ensembles fit and take shape.

## 2.2 A STORY OF TECHNOLOGY

It will be surmized that the history of technology as initially conceived of here is much more ancient than either that of science or philosophy. Its past recedes into the pre-historical eons—perhaps two or more million years—of human and proto-human pasts. And although in recent decades much more knowledge about both the extent and complexity of that past has become known, there is also a certain relative stability to that history until relatively recent times—if millenia can be called "recent."

Today's picture of that past is one which credits our ancestors with a much more complex and diverse set of technologies than once was thought to be the case. The once simple notions of "hunting and gathering" societies have been replaced by the recognition that there are many types of such societies with a diversity of techniques for survival and celebration. Moreover, their intense and specialized forms of knowledge are clearly as complex and culturally developed, although in different ways, as many modern societies. For example, if one takes the knowledge of grains as one aspect of importance, it will be noted that most hunting and gathering societies had a far richer knowledge of such grains than virtually any later sedentary society.

Inland aboriginals in Australia produced over thirty kinds of bread made from more than that many wild grains. Gatherers were more opportunistic with respect to the available resources than sedentary groups—and thus they developed a more complex knowledge about what was edible, and how it could be prepared, than the later more "specialized" societies which domesticated grains for cultivation. We now know that such domestication of grains occurred in many different places of the Earth, with different primary grains accordingly developed—corn in the Americas, rice in Asia and other areas, wheat in the northern lands of Europe, etc. But while some chief or primary grain stood at the center of diet, with maybe half a dozen associated other grains, no agricultural society developed anything like the full range of grain possibilities which nature offered. The surpluses which could be stored in early agricultural society were also much narrower in scope than the natural granery offered. Here was a trade off which in turn was a cultural trajectory leading to what became known as "civilization."

One can also push this comparison much farther. While times of famine and scarcity, to the point of starvation, could and did occur with hunting and gathering societies, dietary monocultures could and did occur in the more specialized societies such that health standards could actually decline or be negatively effected within the advanced society.

All of these changes, however, were slow and often millenia long in development. And if the story of pre-history is today more appreciative of the attainments and complexity of its results than the story told even a few decades ago, there remains the stability of design and use of patterns of technologies over millenia. What we would call innovation apparently was both relatively rare and not valued in the same way as today. But what we now need is a story to parallel our textbook story of science and philosophy.

## 2.3 HISTORIC TECHNOLOGIES

The decisive move from pre-historic to historic eras was, itself, a result of a technology: *writing*. But *writing* in its narrowest, and even possibly Eurocentric sense, had many anticipations by way of intelligible *inscriptions*. I shall take in this history, three variables to illustrate technological revolutions: (a) time technologies, (b) space technologies, and (c) language technologies. I chose these quite deliberately since they are all abstractions which will allow us to relate technological development much more closely to the stories of science and philosophy.

Returning briefly to pre-history embedded in material culture, it is interesting to note that what could be called time technologies occurred very early. Lunar calendars, evidenced in markings on reindeer horns dating back some 35,000 or more years, are already both time technologies and pre-literate inscriptions, or writing. Indeed, when thought of in this way, the contemporary trend to recognize much more complex time technologies in such artifacts as Stonehenge (as an astronomical instrument) and in many other similar devices, throws this unique human-technology capacity itself into pre-history.

Clocks, which are very central to the making of both Western civilization and of science, have at least an ancient historical past. Sundials and water clocks were known to have been invented by the Hellenic Greeks and Romans at the origins of our usual histories. But the mechanical clock did not occur until approximately the eleventh century—indeed, as one of the monumental inventions of the Middle Ages it became a technological revolution.

Natural clocks, of course, are much older in that humans read the sun, the cycles of the solstice and the seasons, the tides, and other phenomena even farther back in our past. The first time technologies were probably attempts to record by inscription, or have reflected within the machinery itself, some natural rhythm. The sundial reads for us the position of the sun, but it also begins to *technologically mediate* that progression. Time, read off the heavens, was at the beginning an astronomical, cosmological phenomenon. Even as late as the early mechanical clocks, that association with the heavens was to maintain itself as most clocks depicted the moon and sun and planets in motion along with the motion of the hands of the clock.

In an early form of runaway technology, as it might be put today, the clock also harbored another possible trajectory. Its regularities, mechanical ones, at first crude, could gradually be perfected into more and more accurate ones, from the earliest clocks which registered only the hours, to later ones with minutes, to still later ones with seconds, down to today's nanoseconds, a progression to finer and finer discriminations or *microstructures,* became possible within the *technological trajectory itself.*

Once this trajectory was taken, at least two other occurrences began to happen. One was the gradual divorce or abstraction of time from its previous astronomical, cosmological context (which usually included astrological elements), and thus mechanical and later forms of time could be taken for their own sakes. And the other was, by returning to a previous cosmological and astronomical context, to discover *anomalies* in the heavens themselves; in effect to *measure the heavens against a now different standard, the technology.* The technology could become the standard against which nature must take account. This, of course, is one of the deep structures of science. It is also both the quantification and the technologization of time. And its roots and basic shape had already been constituted by the time of the rise of Modern Science.

Space technologies are also ancient. Many are related to navigational and travel orientations. Again, inscriptions as a form of prewriting are important. And, as with the natural "clock" of the heavens, reading the traces of animals or other humans in their trails was a high art among our ancestors. To these traces could be added *markers.* The trail could be deliberately, artificially marked for ease and obviousness of its being followed.

Boundary markers, too, are ancient. And from boundary markers which denote territories, there were also property markers.

In ancient Egypt the very foundations of geometry were already founded in the praxis of re-marking fields after the annual floods of the Nile obliterated last year's markers. The Egyptians (and the Babylonians) had a version of the Pythagorean theorum already at hand.

And, just as the clock could gradually emerge from its natural setting, so, also could the space technologies as they became spatial representations in the form of *maps.* The map is to space what the clock is to time. And the trajectories of abstraction, and as a standard against which natural space could be judged also followed. Maps would eventually become, in our history, the world grids of coordinates which idealized and quantified space against which the contingent geographical realities which preceded maps could be judged.

Time and space technologies are deep and broad technologies, technologies which transformed entire cultural perspectives upon the world. But equally transformative in this deep sense, was *writing* which was an early *language technology.* Like time and space technologies, writing, too, had a long and evolving history.

If inscription is the broadest background for writing, forms of record keeping were others. Much early writing is related to bookkeeping, as in cuneiform among the Mesopotamians and the *quipu* cords of the Inca. Writing is one of the best technologies to exemplify the cultural embedment of technologies. Phonetic (from the Phoenicians) writing, in all of its variants in Indo-European languages, contrasts with the ideographic writing of much of Asia (from the Chinese), which again contrasts with sylabaries such as Korean (an invented writing similar to phonetic writing), and all of these contrast with the mixed genres of hieroglyphic, pictographic, etc. Such techno-cultural embedment carries implications to the present where there now exists one trajectory to get the entire computerized world to use a phonetic script, or where technologically advanced cultures such as Japan find different problem sets to develop, as in speech recognition programs that relate to voice rather than writing.

The choices I have made for historic technologies might appear at first to be counter intuitive with respect to the more specific and basic technologies we think of in our histories. Basic technologies, of course, have been referred to in the tools and tool contexts for such basic human praxes as food gathering, preservation and preparation, storage and weaponry, all of which we intuitively place in our usual inventories of technologies. Nor should these be forgotten.

Yet the deep technologies just selected had clearly major impacts upon the development of civilization, and are all examples of technologies as ways of seeing, or perceiving in both a deep and yet culturally embedded sense. They are thus close to the issues which will be discussed relating to technological *problems* in contemporary understanding. They were, moreover, preconditions for the very rise of philosophy and science as we know it.

How this could be the case may be noted by seeing what the three technologies chosen have in common with respect to the *transformation of seeing*. The initial movement entailed in these technologies could be said to be representative, in that the clock, the map, and writing could be taken to represent some other form of activity still enmeshed in more direct modes of action. Natural time rhythms, space arrangements, and speech could be taken as ways in which humans interacted with their immediate environments. By representing these actions in and through an artifact—the clock, the map, and written language—several things could happen at once: (a) that which was represented could be made to "stand still" and frozen, thus it could be returned to and repeated. (b) There was also a distancing made possible by referring now to the standard of the artifact. Here an implicit measuring of the dynamic flow of actual temporality, seasonal changes in the land and sea, and of memory as encoded orally, could come about. The artifact introduces a kind of pre-objectivity in this distancing. And (c) there could also be a "dialectic" or difference introduced such that in that very difference between artifact and natural entity, unexpected phenomena could appear. In this dialectic, in turn, several trajectories could be taken, including different evaluations as to what counted as most important. For example, (d) the sheer repeatability of the standard within the artifact, could be taken as a kind of ideal which nature could only approximate, or, (e) one could begin to see a drift in the differences between the "natural" and the "artificial" (artifactual), in which it would be possible to evaluate one over the other, or the inverse of each.

I am, of course, suggesting here that these three technologies are *instrumental* in more than a metaphorical sense. One could even re-read our previous history of Greek philosophy in the light of these technologies. Could it be that the peculiar valuation of that which stands over that which moves, that that which is distanced over that which is too immediate, and that that which can be ideal takes precedence over that which is bodily arises in the transformations of vision which these technologies produce? This suggestion is, of course, one way of suggesting a line of thought in keeping with the praxis philosophies which originate the philosophy of technology as a latecomer, which nevertheless desires a kind of ontological priority.

This is in no way to suggest that the historic technologies are reducible to these, nor that the incredible variety of technologies should be seen only in these developments. It is to suggest that many of our histories of technologies have, perhaps, been too concentrated upon *productive* (economic) factors. This latter emphasis remains associated with an earlier movement related to technology which I have heretofore not noted.

## 2.4 PROGRESS

The unnoted phenomenon associated with the modern sense of technology is the notion of *progress*. Here the textbook histories—as parallel to the textbook history I followed in the philosophy/science narrative—usually are histories of progress. Many of these are *utopian* in tone.

The utopian thesis is one which usually *measures quantitatively* the phenomena which are supposed to have occurred with respect to the rise of technoscience. It is a narrative which is encapsulated in a quasi-mythology (story) which argues that with the rise of science all previous superstitions, false beliefs, and, sometimes implicitly, sometimes explicitly, religions have been deconstructed. Instead, a fully rational and progressive set of beliefs have come into being through science, which in its applications to technologies, will (eventually) solve most human problems.

Accompanying this narrative are usually extrapolative (and quantitative) illustrations which hold that: (a) life expectancy is much longer now than in any other period of human history, (b) higher standards of living now obtain than at any previous time, (c) food and health standards have dramatically improved, (d) most diseases have been vanquished, and, then, switching to temporal aspects, (e) communication and travel are much faster, (f) knowledge is growing faster, etc. And, although as we shall see, none of these claims is unambiguous, there are elements of truth to each.

Utopian progressivism, reflected in the textbook narrative, is itself a phenomenon of modernity (in the sense of being late Renaissance on). Theories of "ages" have been many. Among the ancients there was a certain popularity to decline narratives or reverse utopias: first comes an Age of Gold, followed by the Age of Silver, followed by the Age of Clay, etc. But our previously mentioned Renaissance humanists (including Roger Bacon, but also many others, such as Sir Thomas More, who wrote the imaginary tale, *Utopia,* thus giving the name to the genre) began to see modernity as the key to a changed and improved human history.

By the eighteenth century, the *Enlightenment* century, progressivism was a virtual secular religion. This period was one in which the rhetoric of "conquering Nature itself" became quite rampant, although its roots were noted as far back as the work of Francis Bacon.

By the nineteenth century, progressivism had become a kind of competing religion. Last year marked the quincentennial of the Columbian voyages to the New World (1992), but Columbus has gone through many transmogrifications. In the quatracentennial period (1892), in a milieu marked by arguments between scientific and religious groups, the story arose that Columbus undertook his daring feat as a virtually solitary individual believing that the Earth was round, whereas his peers and sailors thought it to be flat thus stimulating the superstition that one might "fall off the end of the Earth." This story, which became almost orthodox among school children, was, in fact, invented almost out of whole cloth.

In fact, virtually no educated person in Columbus's time believed that the Earth was flat, although there were two schools of thought regarding the size of the spherical earth—those who still followed the Ptolemaic measurements believed the Earth to be much smaller than it was. And this false belief, held by Columbus, led him to speculate that the islands of Japan must lay only about some three thousand miles west of Spain. He was lucky that the New World turned out to be not too much farther than that! But the tale spread in the nineteenth century was part of what could be called progressivist mythology designed to show that religious people were superstitious; that science dispelled such superstition; and that such knowledge was progressive.

Such utopian progressivism was, of course, not uncontested. Earlier, Francis Bacon, who argued for an interventionist and progressive science which would twist from nature its secrets and form from it the powers needed to build a human utopia, had as his contemporary, Christopher Marlowe, who wrote *Doctor Faustus,* the forerunner of all the mad scientists and Frankensteins who, for the sake of a kind of technological power, made a pact with the devil. And the later Enlightenment was to find its dystopian detractors in the nature idolization of the European Romantics. The literary and rhetorical battles of the utopians versus the dystopians have long been a part of modernity itself. They also underlie many of the arguments and problems associated with technology.

Progress, however, lies deeply embedded in the culture of science. Knowledge is thought to be progressive, accumulative, and qualitatively on a trajectory of either evolutionary or revolutionary improvement. And insofar as technologies are today associated with science and its culture, the same belief in progressivism is often held. That is why contemporary doubts about technological progress cut so deeply.

There is even a kind of irony in some of the contemporary debate concerning technology, in that there have emerged what could be called negative progressivists. These are the dystopian thinkers who have *adapted the quantitative tactics* of the utopian progressivists, and argued that the march of technology has, indeed, been a quantitative extrapolation, but a negative one. There is now demonstrably more pollution, more environmental degradation, greater human woe overall, precisely due to technological progress. Such negative progressivists accept the extrapolated change model of modernism, but invert its values. All of this makes for great difficulty in a careful, balanced, and *critical* analysis.

Such a critical analysis implies an *assessment* of that related to technology which we wish to examine. Progressivism, in both its utopian and dystopian modes, often reverts to a quantitative analysis for its arguments. But this very mode of argument is itself associated with the rise of modernity as well. In philosophy the attempt to quantitatively evaluate can be associated with the late eighteenth, early nineteenth rise of *utilitarianism* such as that conceived of by Jeremy Bentham (1748–1832). His argument that ethics should be based upon the principle of the greatest good for the greatest number was bolstered by attempts to *quantify* such goods. In our history of technology, it should not go unnoted that this mode of argumentation arises in the very midst of the industrial developments which stand as predecessors for the twentieth century. These arguments will again appear in the problem of technology assessment which has become a contemporary problem as well. What can be said is that such a calculus of goods (not *the* Good) is an analytic approach to ethics. It, like a technical process, must somehow measure or assign quantitative values to human goods. A quantitative utilitarianism anticipates the notion of risk analysis.

The question at this preliminary high altitude would be one of how one measures progress, whether positively or negatively. It should be clear that the occupants of extreme positions frequently leave unqualified the claims made: it clearly is true that, at least in industrially advanced countries, more people live beyond early childhood and have health technologies which prevent many premature deaths. But while more people living closer to the human biological horizon lifts "life expectancy" figures, the horizon of human life today is not different from ancient times. It is true that some diseases have been totally eliminated (smallpox), and others greatly reduced (tuberculosis), but there are also new diseases (HIV or AIDS) and others that are spreading (malaria) due to selective resistance brought about precisely by our control mechanisms. And while we have a clear and general sense that health standards—again in industrially developed countries—are higher than those of the past, the question of whole earth measurements is unclear.

For the negative progress quantifiers, one can note that as bad as contemporary urban smog is now, it was presumably considerably worse in the early days of the Industrial Revolution when coal fires were the standards of home and industrial heating. And, once river systems are cleaned up and fish ladders reinstalled, many twentieth-century river systems show fish which had been absent for several centuries again returning (the Thames in London and the reintroduced salmon runs in both European and American streams). Lead poisoning from pewter dinner plates, common in our North American beginnings, is now virtually unheard of. But, of course, PCBs were also unheard of in the eighteenth century! In short, even with the best quantificational methods, it is not unambiguously clear what, overall, is better and what worse. The question remains: how are the effects of technologies to be assessed?

The quantifiers may be correct in that there is "more of everything": people, information, physicists and philosophers and engineers, communication, etc. But this, too, is ambiguous without some qualitative way of analysing the result. Moreover, progress itself is a cultural sediment, perhaps uniquely related to our Euro-American histories which also must be placed in context in any philosophy of technology.

## 2.5 TECHNOLOGY AND CULTURE

I have suggested that all technologies are embedded in culture. Insofar as they are human products that should be intuitively obvious. But insofar as technologies entail "ways of seeing"—whether or not Technology (with the capital) equals the metaphysical way of seeing (Heidegger)—that, too, implicates the particularities of culture.

In this light it is preliminarily important to note that in the broader history of technologies, there have been some very different sources and times for technological development, not all of which fall into our own tendency to take our history as unique.

Lynn White, Jr., in his analysis of the technological revolution of the Middle Ages, has suggested that the Latin West has had a peculiar penchant for technology-as-power—and that this tendency existed even before Francis Bacon made a virtue of it.

Joseph Needham has, similarly, undertaken careful studies of a different history of technologies in Chinese civilization. Almost everyone is aware that the compass, explosives (later, gunpowder) rockets, balanced rudders, and waterproof compartments on ships, and many other devices were of Chinese origin. And while many of these devices had actual practical uses, many others were associated with *celebratory* praxes. One might even engage in a bit of imaginative history by noting that whereas the European inventors of the telescope immediately sought to sell it as a potentially useful piece of military hardware to the princes of the day, the Chinese use of powder for firecrackers and rockets for fireworks were first invented and put to use in festival settings. Later, however, they too reverted to military uses and were the forerunners of cannon, artillery, and other destructive devices.

Yet even the playful use of technologies could display a different kind of cultural chauvinism, as Daniel Boorstin has reminded us. In one period of Chinese cultural and political expansion—ironically only a few years prior to Columbus's voyages to the New World—the emperor, Yung Lo, sent his admiral, Cheng Ho forth to colonize the Pacific and Indian Oceans with a fleet of ships, of which the smallest was nearly double the size of Columbus's largest. His weapons were highly crafted gifts to be *given* to each of the local potentates found as a *declaration of the superiority of Chinese culture.*[1]

Such very different patterns of cultural technological embedment, however, leads to another feature in the history of technology which must also be noted. *Cross-cultural exchanges have frequently been the harbingers of high technological innovation and development.* I have previously noted this phenomenon in passing within European history. The Hellenic-Roman periods of strong engineering and technological development were periods of strong cultural and trade interchange within the African, Oriental, and Mediterranean worlds. Similarly, the centuries following the awareness of the Far East after Marco Polo, among which was the European craving for spices, brought in its wake many of the crosscultural exchanges which White has noted in the building of Medieval Europe (arches, stirrup, windmills, etc.).

And, if this is a past historical motor for the rise of different and differently utilized technologies, one should not miss the contemporary version of the same phenomenon. The movement of Western technologies outward seems both obvious and the dominant movement of crosscultural technological dispersion. But, as we shall see, it is not the only dynamic part of the crossculturality implicit in technological history.

# CHAPTER THREE

# PROBLEMS AND TECHNOLOGIES

We now have before us two tales, one of the history of philosophy in its early relation to science and its more recent relation to technology, and one of at least an incidental story of technologies, particularly as they have related to cultures and, in our Western context, to science/philosophy.

It may seem, correctly, that the relationships between these two histories are often far from direct and even that there are frequent disjunctions, particularly between the high moments of theoretical advance and the explosions of technological innovations. And, were we to believe the standard textbook histories which may also be called attempts to make history orthodox as in a rule of true belief, it might well be claimed that while this set of disjunctions has obtained for much of pre-Modern history, at least since Modern Technology, the disjuctions no longer hold. Modern Technology, presumably, is *scientifically derived* technology, or applied science. The insight in this belief is one which recognizes that science in its contemporary mode is very closely linked to technology—but perhaps not in the way in which the orthodox history holds.

This chapter will begin to introduce *problems of technologies.* And I put it this way because I do not hold that there is one Problem of Technology (with capitals) as did some of the early dystopian philosophers of technology. Two such "problems" have already been alluded to, one close up for philosophy in its connection to the history of science, a problem I shall term a *micro-problem,* and one a *macro-problem,* that of the overall impact of technological processes as a geological force upon the globe, the environmental problem.

Between these two problems lie a whole series of other, middle-sized problems. Many of these relate to the very recent emergence of hi-technologies. These run the gamut from the impact of computers on everything from artificial intelligence, to VDT (video display terminals) and micro-radiation, to the questions of networks and civil privacy rights, and also to hi-tech medicine which changes the very parameters of life-beginnings and life-endings, as well as what happens in-between. Insofar as these problems have been addressed by philosophers, three traditional interests of philosophers have dominated: epistemological ones regarding what and how we can know things, and ethical ones regarding what we should or ought to do, and socio-political ones regarding the structures of society. Indeed, in much of contemporary Anglo-American or analytic philosophy, the preoccupation with computer models and impacts for philosophy of mind is a major theme. And, in medical ethics, often dominated by followers of utilitarian traditions, the role of hi-tech medicine is held to have created a whole series of new problems with respect to calculating human rights and goods. And, among neo-Marxists and Critical Theorists, the question of social and economic justice is raised vis-à-vis technology.

A second set of problems related to technologies come closer to the macro-problem of technology and the environment, but are more specific with respect to areas of concentration. These include the impact and problems associated with Big or Corporate Technologies, such as agribusiness, the petro-chemical industries, the defense industry, and multinationals. Here the philosophical focus is one which usually revolves around political and social philosophy. What are the forms of society and the possibilities for the just distribution of goods and social justice? In these areas our neo-Marxian thinkers, along with liberal democratic traditions, have had a good deal to say.

Another set of problems related to technologies, might be called socio-psychological or humanistic in focus. Here I refer to the effects of technologies and the human contexts which change because of these. Is technology alienating? Dehumanizing? Or is it fulfilling? And choice multiplying? One whole strand of argumentation about technology has arisen in philosophy between those who see in technological development as a diminution of many highly valued traditions, usually associated with community and scale, and those who see in the same developments, the growth of pluralism, cosmopolitanism, and a postmodern context.

As I have suggested, the literature on these problems has proliferated in the twentieth century, particularly since the middle decades, and it will be impossible to address any but a selected set of problems. I shall do this by first addressing a problem which lies closest to the philosophers who have addressed science and technology questions directly from the Anglo-Euro-American traditions: the science/technology question. I shall then selectively deal with a couple of middle-sized questions to indicate something of their parameters. And, finally, I shall return to the macro-problem concerning the ensemble of technologies and the impact upon the global environment. It should be noted that in this chapter I shall be dealing with issues which have been and are currently being discussed, particularly within the broad and often interdisciplinary context in which philosophy of technology moves.

## 3.1 SCIENCE/TECHNOLOGY

If there is a problem which has concerned most twentieth century philosophers of technology, and almost all philosophers of science interested in any way in technology, it is that of the relationship and difference between science and technology. This is a close-up problem for philosophy because it entails the *concepts of both science and technology* and obviously implicates the much larger problems about theory and practice, problems which exceed those of science and technology per se.

Moreover, this problem is one which is basic, even though narrow. It is important that we get our notions of science and technology right if we are to follow out the implications of either or both for the contemporary condition of humanity. One can see how this works if we ascend to a very high altitude and relate what is implied in this question to the eventual macro-question of technology and the global environment.

Example: *if* modern (theoretical) science is, indeed, the motor which drives and develops technologies in the forms of applications, and if these applications are what lead to and effect the global impact of technologies, then because science is basic to this development, it would seem that the only way in which we could solve our environmental problem is to do so at the source: *science and its entire theoretical context.*

There are philosophers who take exactly this tactic, preeminent among them have been some recent *feminists,* who often come closest to attacking precisely the cultural context of Western science itself. This form of critique is, itself, largely a late twentieth-century occurrence.

Although various arguments are made, sometimes leading from the neglect science as an institution has had for women, both by forgetting their actual role in the development of science and by the poor representation of women within (particularly natural and engineering) sciences, to a much larger claim about the *paternalist model implied in science-as-institution.* Here our previous history is rejoined and retold, particularly in its Baconian moments.

Sandra Harding, one of the leading feminist philosophers of science, argues in her book *The Science Question in Feminism:*

> When Copernican theory replaced the earth-centered universe with a sun-centered universe, it also replaced a woman-centered universe with a man-centered one. For Renaissance and earlier thought within an organic conception of nature, the sun was associated with manliness and the earth with two opposing aspects of womanliness. Nature, and especially the earth, was identified on the one hand with a nurturing mother…and on the other hand with the 'wild and uncontrollable [female] nature…. In the new Copernican theory the womanly earth, which had been God's special creation for man's nurturance, became just one tiny, externally moved planet circling in an insignificant orbit around the masculine sun.[1]

In short, the rise of early Modern Science was itself a movement into the Baconian, masculinist context of an aggression upon nature betrayed in the metaphors of science "twisting the tail of nature" or even the use of rape metaphors which proceeded from Bacon on into very contemporary speeches by Nobel Prize acceptees in the last decade.

Of course such critiques take science in a certain way—science-as-historical institution—and not as a mere instantiation of "pure theory" in an ideal form. Theory, in the feminist critique, is historically institutionally contexted. As such, feminist philosophers of science—and they have become major contenders among contemporary philosophy of science—are part of another chapter in our narrative of the history of science and philosophy.

This suggested line of critique arising out of feminism, is a *radical* re-conceptualization of the history of science/philosophy. Another such radical re-conceptualization was suggested at the origins of the philosophy of technology. Both Dewey and Heidegger, in effect, argued that the usual conceptualization of technology being derived from science is wrong. Rather, science should be seen as a necessary tool for the development and operation of technology. But here, Dewey and Heidegger took different tacks. Dewey, taking instrumentalism as a positive mode of inquiry, argued that there were essentially instrumental praxes involved with both science and technology. Both are pragmatic modes of inquiry, and science in its highest theoretical moments is also a reflection of practice and instrumental ends. And Heidegger, in a consciously non-gendered way, anticipated some of the feminist critique. He argued that implicit in the Western concept of nature *viewed technologically,* was a nature solely there as a resource well or standing reserve of energy for human (masculine?) use. Such a technology must utilize the exact methods of science to successfully attain this dominating goal. Thus Heidegger, like Harding, saw in our tradition a certain way of seeing/taking nature, but he lays this conception at the heart of technology. All three of these views, however, entail interpretations of both science and technology, although they cut the issues very differently than the textbook history which remains dominant, though which is coming very much under attack.

Most textbook history remains one in which an ideal theory procedure remains the heart of science. Its result is supposed to be the establishment of an objective, value-free, and unfettered truth which arises from a hypothetical-deductive process. Such a view which often ignores the constructive and culturally-laden aspects of science is a legacy of both late-nineteenth century and earlier Positivist views of science.

Were this view to be the correct one, then one could argue that free and unfettered (theoretical) science is itself value neutral and only when its theories are applied can good and bad uses come of these. This means, of course, that only at the applied level can questions of value, blame or praise, come into play, which in turn means that problems of technology are simply problems *of* the application of science, not of science per se. I have indicated that such a view has come under increased attack in the mid- to late-twentieth century, but for this we must return to our tale of the philosophy of science, now as it increasingly relates to the philosophy of technology.

## 3.2 SCIENCE'S TECHNOLOGIES

Even the textbook histories acknowledge that early Modern Science, in contrast with its Greek origins, was *experimental.* But to be experimental, in the historical context of early Modern Science, meant that (a) an experiment *used technologies, i.e., instruments,* and (b) was placed in a situation in which the natural phenomenon was controlled or put under certain constraints. The social construction of this latter specific way of addressing, or even manipulating nature has been well addressed in the standard histories. But the philosophy of science, with its theory-bias, has been negligent in emphasizing the technological nature of the first element of experimentality.

That imbalance was to be corrected by the late-twentieth century. I left the story of North American philosophy of science at its inception and the coming to dominance of largely Positivist strands of science interpretation. But the story takes a very different twist from the 1960s on. In 1962, Thomas Kuhn published *The Structure of Scientific Revolutions,* itself to be a self-fulfilling revolution in the philosophy of science.

He argued that actual science, in contrast to the then textbook history science, often underwent dramatic *revolutions* in which whole new ways of conceiving of things replaced earlier notions. Such revolutions were gestalt shifts which were cultural-theoretical ways of seeing. Thus, as Kuhn put it, Aristotle seeing a pendulum would have seen it as an instance of restrained fall, whereas Galileo would have seen it as an instance of momentum and the laws of inertia. Kuhn's book today remains the single most cited work by humanities writers interested in science and his use of paradigm shifts has entered common parlance.

His work, however, was but one in what soon became a flood of important books reinterpreting science and its *praxis.* Karl Popper in England attacked the leading Positivist interpretations regarding theory verification, and moved to an interpretation of science both as institution embodied in research programs, and as limited to a theory progression dealing, at best, with falsification procedures. Imre Lakotos pushed the Popperian theories farther, and in the United States, Paul Feyerabend began to argue that there was nothing intrinsically or intuitively superior to science's theorizing than other modes of theorizing, in *Against Method* (1975). The 70s and "post-Kuhn" philosophy of science began a much more pragmatic, historical, and problem or region-oriented approach to science. Even within the dominant center of mainstream (analytic) philosophy, there had been attacks upon the older distinctions between synthetic and analytic statements and upon the differences between observation and theory-oriented propositions—W.V.O. Quine in a Deweyan-inspired form of analytic pragmatism often stood out here. And like the earlier Deweyan stance, the import was again to diminish the differences between both science and philosophy, but implicitly also between the previous neatness assumed to underlie science and technology in its contemporary guise.

If these directions began to take back from science some of the presumed theory eminence granted it in the previous decades, another part of the proliferation of a pluralism of philosophies of science was to drive that sub-discipline closer to questions of technology. There emerged in the 70s and 80s what I have called a group of instrumental realists regarding science and its essential technologies.

My 1979 book, *Technics and Praxis* was part of this movement. In that book I argued that Modern Science, in contrast to its ancient and more contemplative origins, was *essentially and necessarily embodied in technologies, instruments.*

The core chapters of *Technics and Praxis* were devoted to a phenomenology of instrumentation to show how science is embodied in technologies, but more, how those technologies in their very non-neutrality played essential roles in trajectories of inquiry at the center of modern and contemporary scientific knowledge gathering.

Then in 1983, two books appeared independently and simultaneously, which followed a similar instrumental realist interpretation: Ian Hacking's *Representing and Intervening,* and Patrick Heelan's, *Space Perception and the Philosophy of Science.* Hacking's book was divided into two parts, the first of which might be thought of as more traditional philosophy of science insofar as the problems addressed were theoretical-conceptual ones addressed in part to the realist and anti-realist controversies that are still contemporary. But in the second part of the book, Hacking reminds his readers that early Modern Science at its origins was Baconian, and far from a mere theorizing, was an instrumental (read, technological) *intervening* with nature. He illustrates this particularly well in his history and philosophy of microscopy. Hacking clearly saw the essential role played by science's technologies.

Heelan, who had argued the point in articles as early as the late 60s, directly addressed science as a technology-embodied enterprise in which instruments are readable technologies through which the realities of previously unperceived entities come into perception itself. He even takes the case so far as to say that a scientifically constituted object *can only be what it is by being instrumentally or technologically so constituted.*

Two years later, but again independently, Robert Ackermann

published, *Data, Instruments, and Theory* (1985), in which he argued that what counts as scientific progress and the accumulation of knowledge which belongs to the scientific enterprise, is constituted through the improvement of the instrumentarium, again, science's technologies.

These instrumental realists not only redressed an imbalance of interpretation concerning science's relation to technologies, but saw in instrumentation *an essential feature of science.* This factor was taken even further by other philosophers in the late 80s and up to the present. Peter Galison, a historian-philosopher, saw not only that science is technologically embodied in the experiment, but that the complexes of instruments in science constitute their own kind of instrumental traditions, especially in his book *How Experiments End* (1987). But most radically of all, Bruno Latour in *Science in Action* (1987), argued that not only is science so instrumentally embodied—particularly in the laboratory which supercedes the observatory paradigm of earlier, more purely observational, science—but that at the very heart of the experiment and in the apparatus, there is the constitution of scientific objects through the process of *inscription. Laboratory machinery is, in effect, a complex writing instrument.* We are thus returned to another variant upon the deep technologies previously alluded to. Latour sees in this close relationship of science and technology, something very akin to the previous inversion of the science/technology relation which arose with the origins of philosophy of technology. He terms contemporary science, *technoscience,* a term which is now gaining more common usage.[2]

We may now return, once again, to our story regarding the rise of early Modern Science. At its inception in the Renaissance, its recovery of the ancient nature philosophies was one which fit immediately into the engineering emphasis of the sixteenth century. Da Vinci's fabulous machines, Galileo's telescope and inclined planes, and Bacon's instrumentarium, were all part of that first dimension of experimental science, science's essential technologies.

Two other dimensions of this binding of science to technology at its inception should also be noted: first, the underside of the need to have technologies, is the need to *fund them.* And if the Renaissance was engineering-minded, it was also *opportunistic.* Machiavelli reflected the entrepreneurial politics of the day in his book *The Prince,* but these politics were also followed by our patron saints of early Modern Science. Da Vinci wrote letters to the Duke Il Moro, arguing that he should be patronized by that duke because he could prove useful in inventing many instruments of war. Galileo wrote the same kinds of letters to the Medici family. And, much later, Werner Heisenberg was to attempt to save the new physics by appealing to Hitler concerning the technical possibilities of atomic energy. Institutionally, early Modern Science was not neutral or value-free.

Second, although less dark, but equally implicated with technology was a way of seeing which the high Medieval lifeworld provided for the Renaissance, that is, a mechanically metaphored and technologically constructed lifeworld. Principles of ballistics related to warfare, of a geometrized space in the laws of perspective, of a gridded world mapping in the School of Henry the Navigator, were all part of the way the world was seen at the rise of early Modern Science. Such was a modern, technologized world.

All of this is meant to show that from the beginnings of early Modern Science, science was already *technologically embodied.* This has been increasingly recognized now by a whole spectrum of what could be called new philosophies of science. These include feminist, instrumental realist, pragmatist, hermeneutic, and social constructionist analysts of science. What is common to this set of thinkers is the thinking of science as a context-bound social and cultural institution. Science is a *specialized praxis* among other such praxes of human activity.

Moreover, it is clear that there is *some deep and essential bond between science and technology, without which Modern Science would not be what it is.* All of this belongs to science even before it became Big Science. It should be noted that the now popular term, Big Science, was also part of the revolution that began to take place in the 60s. Derek de Solla Price—who also saw in crafting of instruments much of the success story behind science—wrote *Big*

*Science, Little Science* (1962), for he saw in contemporary science yet another step in the science/technology relation, the emergence of a *corporate structured science*.

I have also alluded to this development which followed the Industrial Revolution of the nineteenth century and the World Wars of the twentieth century. A corporate science, however, is solidified only in the twentieth century. Symptoms of this phenomenon can be seen in the way in which the leading sciences have operated and set models for other sciences to follow:

1) There has clearly been a major change in the way in which the individual genius scientist pioneering some new discovery in a small and isolated laboratory, finds a breakthrough and leads the world to his/her discovery. This has been the romanticized story which still is often portrayed as a history of discovery, and which had some truth with respect to the past. Today, the discovery is much more likely to be made in the context of a large and complexly organized and funded laboratory—probably costing millions of dollars—staffed by a factory of technical assistants, post-doctorals, administrative assistants, and a small number of name-recognizable senior researchers who spend as much time publicizing and campaigning for new grants as they spend time in the lab itself.

2) Authorship changes are also symptoms of Big Science. *The New England Journal of Medicine*, one of the most prestigious of health science journals, reports that the 95 percent individual authored/5 percent jointly authored ratio at the beginning of this century, has now been reversed with approximately the same ratio, that is, now 95 percent jointly authored articles.

3) The biggest and most complex Big Disciplines frequently dominate and set the model for the smaller disciplines to follow. Of these, physics is clearly the largest and most successful. Its current origins are related to the Manhattan Project (atomic bomb), and it is interesting to note that a physicist-turned-biologist is a leading figure in arguing for and getting funding for a human genome project which will be biology's first more than billion dollar project!

4) Instruments, which as we have seen are the embodiments of science, now have reached the multibillion dollar level. The Hubble

telescope, a small and only few billion dollar instrument; the super collider, a large and complex single instrument; and the science/engineering proposed space station at well over fifty billion, are examples of complex, corporate *technoscience* in the late-twentieth century.

5) Moreover, the institutions within which technoscience operates are multiple and overlapping. Today they include universities (primarily the large research universities which hold over half the total faculty and student bodies related to science), corporate industries (including the new ones, such as the computer and biotech ones spun off from university groups), government agencies (such as the National Institutes of Health), the military (funding for science research ranges from approximately 35 percent Department of Defense funding for mathematics, to over 85 percent for aeronautical engineering), and private foundations (including the Hughes, Ford, Exxon and other such groups). The picture noted in 1) is very hard to find amid this contemporary picture of technoscience in practice.

It is also easy to see that technoscience is *more* an intermixture of science and technology than ever before. Thus to speak of science in the contemporary world is ipso-facto to also speak of technology. This observation is not intended either to blur or eliminate the problems which continue to be discussed concerning the science/technology relation. It is intended to see these arguments, however basic, as properly scaled as micro-problems with respect to other issues of equal or even greater importance with respect to understanding the phenomenon of technology.

## 3.3 TECHNOLOGY BEYOND SCIENCE

It is clear, from the preceding discussion that much of contemporary technology is science-related. And if a rather conservative view of technoscience is that science has moved into technologies in various ways, the more radical tendency is to have reinterpreted science as both guided by, and necessary to the trajectories implicit in technologies. But we have also seen that many historical technologies have been developed without science and many continue to be developed without explicit ties to science, institutionally.

In contrast to the previously noted disjunctions between theory periods and technology periods in our past history, it might seem that today most technologies are (a) more closely related to science processes, and (b) that there are deep and essential differences between modern technologies and their prior traditional forms. Today's electronics, computer technologies, space technologies, and most hi-tech products are clearly technoscientific and could not have been produced without some deep relation to science, however conceived. And some philosophers have emphasized this break in the history and type of technologies.

How this is so may be illustrated nicely by the invention of *plastics,* clearly a modern material *par excellence*. Plastic is not found in raw nature. And it is even more extremely different from the materials used in ancient technologies than might first appear. I have previously referred to found technologies, the stone or stick or gourd simply picked up and brought into human praxical use.

The next step up was one which took a found product and modified it in some way to make its use better. One removed the twig nubs from the branch or smoothed the edges of the broken gourd, thus modifying what was found, but the material itself remained that which presented itself to such uses. It took humankind a very long time indeed to achieve yet another step—the modification of the materials themselves.

The invention/discovery of such material possibilities as glass (from melted silica), and pottery, and most important, the forging of metals (bronze, iron, and much later, steel), began to make the transformation of technologies occur at the molecular level, but without a theory of the molecular. All of this occurred well before alchemy!

Plastics, however, take the process a fairly large step farther, in that plastics—much more unlike the metals and metal compounds or glass, which can be found in nature—do not so occur there, but are manufactured through precisely the processes made possible by technoscience. The long strings of polymers are a constituting of a new material through invented molecules or molecular arrangements.

And the temptation might well be to revert here to the theory-bias once again and note that only when we have a sufficiently sophisticated theory development can plastics be made. That is true, but it is equally true that the capacity to perceive at the molecular level is precisely the result of technologies—instruments—which bring the micro-world into scientific presence. Plastics rather perfectly illustrate the possibilities of *techno*science.

Contemporarily, the search for new, human-initated materials constitutes a major thrust of technoscience, and occurs in the whole spectrum of hosting institutions from universities to corporations. In principle, this capacity to manipulate the molecular is no different for technoscience in materials research, than in biotechnology with its still only beginning techniques of DNA or gene manipulations and manufacturing. Here, again, is a Baconian technoscience moving into new areas for the constitution of new biotechnological entities which will be conceived to serve human purposes according to their inventors. What is different contemporarily in contrast to the traditional technologies is that the micro-level at which such manipulations may occur has now been lowered to the molecular, even atomic level. Indeed, one of the first public illustrations of single-atom manipulation was the arrangement of thirty-five atoms to spell out the IBM logo! That is part of the greater magnificational/amplificational power of contemporary technoscience over any of its historic precedents.

What is also revealed in this perspective, is the way in which a technical way of seeing is often implicated in all technoscience developments. What is possible is defined by the ability to manipulate entities, and this, in turn, by the increasing range—both micro and macro phenomenal—of the technoscientific instrumentation.

These new powers, applied in ever more areas, are also much of the source which get identified as problems of technology. I propose here to look briefly at two which have been addressed by philosophers:

1) *Life beginnings and life endings.* One of the earliest areas of

applied philosophy to receive attention and to begin to interest philosophers in technologies related to rapidly developing hi-tech medicine. As new technologies began to be available—usually with very complex and expensive machinery and thus with limited applicability—the medical profession began to worry about decision-making in the application of these therapies. Early kidney dialysis machines were among the first to receive attention.

The ethical situation which arose revolved around questions of who, considering limited machines at high costs, should receive the benefits. Physicians, not always reluctant to play god-roles, began to have doubts about bearing sole responsibility for such choices, and hospitals and medical schools began to develop various ethics advisory panels—often these included professional ethicians such as philosophers and theologians. The famous Hastings Institute which began its work in medical ethics beginning in the 1970s, often studied such problems.

I shall not go deeply into the history of this development, but it should be easy to see that questions about who deserved to be high on the treatment priority list followed. Someone of great social worth? Anyone who could afford to pay the high fees? A random and thus fair lottery? This scarcity situation was, of course, well known in philosophical ethics. It is often called the "lifeboat situation": who gets sacrificed in order that some survive? Nor did this problem relate only to kidney dialysis—it applies to transplanted organs, to artificial hearts, to many, many contemporary medical techniques.

In the contemporary institutional setting it should be noted that many of the parameters involving such increasingly difficult situations were problems which arose in relation to expertise and values. Medical experts are not necessarily ethical experts, and many times what was desired was, in fact, to find such an ethical expert. This was precisely because, once again, the technical environment was such that were a solution to be possibly modeled upon technical problem solving, it would seem to be more easily resolvable. This, in turn, made for an environment within medical institutions which favored *ethicians who best fit the mold of calculating costs and ben-*

*efits on the basis of ethical or social goods.* It was not surprising, therefore, to find that various brands of *utilitarians* often filled these positions. I have already alluded to the predisposition of utilitarianism toward the early industrialization model of late modern philosophy in the nineteenth century in my discussion of Jeremy Bentham.

On a deeper level, however, philosophers also began to realize that hi-tech medicine was implicitly redefining what could count as both life beginnings and life endings. What is changing might be characterized as a boundary between what is spontaneous and what is decided. Crudely, it was not too long ago when the stoppage of breathing, or a heartbeat, or of other vital signs was taken simply as the end of life after which the patient or victim was simply dead. Today, at least in industrially developed countries, none of these count, *necessarily,* as life endings. CPR, resuscitation, fibrilation, and many other technologies and technique processes can bring the patient "back from death." One simply does not let nature take its course—*except deliberately.* Thus arise the complicated ethical questions which are virtually publicized daily: living wills, when to remove life-sustaining technologies (thereby involving some decision about what shall count as life), and other life ending processes. One could call this a Sartrean dilemma after Jean-Paul Sartre's famous existentialist notion that one must always decide one's freedom since one is condemned to freedom. Today one must decide one's death which brings the whole issue of ending life very close to the question of choosing, in a nuanced way, suicide. It is not accidental that rational suicide has become a major topic of discussion by ethicians, philosophical or not.

The same structure may be found in life-beginning phenomena. It has often been claimed that much of the feminist revolution of the twentieth century could at least be indirectly linked to birth control technologies and techniques which were genuinely effective, thereby giving women greater control over their own bodies. But whether that is or is not the case, what is clear is that, again for highly-developed technological societies, there is something of an inversion about what counts for sexual and reproductive decision and responsibility. Today, one decides to reproduce much more often than in pre-control eras. One might even be considered irresponsible not to take rational account of prevention or planning. Similarly, the variety of techniques and technology-related choices facing the contemporary couple has been multiplied in ways exactly paralleling the life-ending proliferation of means. In vitro fertilization, now successful and popular, is but one complex process which, like spontaneous death, does not leave nature alone.

Hi-tech medical technologies present us with a symptomatic middle-sized problem indicative of many other such problems relating to contemporary technologies. Clearly such technologies *simultaneously provide us with more choices and alternatives than we had previously, but while so doing, there is a much more intense burden placed upon consciousness and deliberate decision making.* Again, in a Sartrean mode, one is condemned to play God. Nor can one avoid the technological context which shapes and situates this contemporary problem set.

2) A second middle-sized problem increasingly discussed by philosophers interested in technologies, is one which relates to *hitech agriculture.* This problem combines, not dissimilarly to the above, revolutionary twentieth-century technological techniques with large institutional matrices—agribusiness.

It is also a problem which follows a trajectory of initially utopian hopes, leading to critical and sometimes disillusioning worries, to an appreciation of the deep ambiguity facing technologically advanced cultures.

We have already noted that the roots of agriculture, both domestication of plants and of animals, lie in pre-history. Selective planting of larger seeds of a selected variety slowly led to seed hypertrophy (enlargement) to the point that it became not only hard to distinguish the domestic plant from its wild predecessor, but domestication often implied human intervention such that neither the plant nor animal, once domesticated, could easily return to nature. Even the Old Testament's Jacob used selective breeding and fooled his father-in-law (Genesis 30:31–42).

Yet much of this technique was not immediately technologized.

The modern mechanization of farming was to arrive quite late. The plow had revolutionized feudal praxis, and mills were used to produce food stuffs, but all these were relatively lo-technologies. This is not to demean the accomplishments of the agricultural revolution. Central Europe, still the most populated part of the globe when person per acreage is counted, learned to more than supply needs. Peasants even learned to build up soils through composting and recycling. But much was done by hand or with oxen or horse.

Mechanized agriculture, even with machinery of relatively simple-geared and machine-driven technologies, was relatively late in arrival when compared to urban and industrial development. Reapers, at first horse drawn, planting machines, and almost all sorts of mechanical picking devices appeared only in the late-nineteenth and early twentieth centuries. Today, of course, there are combines for all grains, and even for picking grapes for wine production. Moreover, jokes are often made about attempts to breed square tomatoes to fit boxes for current transportation technologies.

The modernization of agriculture, interestingly, also fits into the early scientific revolution in other ways. In the eighteenth century, breeding practices began to be undertaken by specialists for the first time, and only in the nineteenth century with the rediscovery of Mendel, did a theory-oriented version of selection and breeding begin to occur as a kind of practical genetics. Hybrids, then double-cross hybrids, began to be constructed only after agriculture had already become agribusiness.

Modern agriculture is, of course, tied into modern problems, including that of feeding the rapidly expanding population. The basic axiom has been that ever more food is needed to supply the ever-growing population, which, itself is a product of technoculture. The population explosion is the indirect result of lowering infant mortality and bringing more people toward the life horizons through hygenic and modern medical practices. (Birth rates, however, once a country is highly industrialized, tend to themselves fall into a different and technologically textured pattern. Japan, which had approximately five children per family at the end of World War II, has, like all such countries, declined to approximately two children per family today. China, far less advanced, is trying to emulate this number with a more centralized set of practices.) Agriculture has often been charged with keeping up with this human population expansion. And this need has been part of the motivation to meld agriculture into a more technologically intensive, efficient, and productive enterprise. In the process, the well known Green Revolution which was based upon engineered hybrids, monocultural planting patterns, and the uses of agrichemical products (herbicides and pesticides) expanded.

However, in 1962, Rachel Carson's book, *The Silent Spring* appeared. She was one of the first of the critical writers from within science to sound an alarm, and she showed that one effect of hi-tech agriculture had severely negative effects. Pesticides (particularly DDT) were demonstrated to be destructive to various forms of animal life and possibly damaging over the long term to overall health standards through the toxic residues left in plants and the soil. This warning has today escalated to highly magnified health concerns about residues across a very wide spectrum and plays an important role in the discussion of whether hi-tech agribusiness is desireable, and if not, what modifications and substitutes might be found to maintain the world's food supplies.

The macro-outline of modern agribusiness is familiar. As a system it parallels precisely the development of modern technoscience and often forms, in the process, another dimension of the contemporary environmental problem. Let us look at but two sub-problems in passing: (a) *Monoculturality.* Nature, conceived of as a resource well according to Heidegger, needs its resources managed, and managed efficiently. This is part of the rationale behind much monoculture development. A stand of pine trees, pure and isolated from any other intrusive plantings, is a pulp resource for paper production. Their uniform growth, spatial (in rows) arrangement, and harvesting convenience, all are part of the planned efficiency. Similarly, biotechnologically managed breeding—artificial insemination with a single male animal of presumed best quality characteristics—would presumably produce the ideal milk herd. The price of this monoculture, of course, relates in different ways to *diversity.*

In the first example, a monocultural forest is habitable only by a very restricted number of animals, thus it reduces the natural diversity which might inhabit a previously diverse tree culture. Or, in the monocultural breeding program, any unforeseen weakness—such as showed up in Australia in 1990 with a double recessive gene problem which caused one-quarter of the calves to die early—quickly propagates through the herd.

We border here on the environmental problem, and philosophers looking at this system often wonder about alternative models of human-nature relations. Here the various alternative models of appropriate technology, deep ecology, and managed diversity all appear.

(b) Food production remains central to agriculture, but the *technological transformation* of food is also instructive. It is, of course, easy to point to fast food and its proliferation around the world as one product of technoculture. But almost all food processes imply some degree of technological transformation. Cooking is such a transformation, but much cooking is minimalist. That is the case with stir-fry, with steaming, and even the barbecue, in each of which cases it is essential that the ingredients be fresh and very little modified.

As Lawrence Busch has pointed out, technological agriculture underwent a much more maximalist transformation in the early days of the twentieth century. Food stuffs, canned and preserved, were transformed from fresh products to highly preserved or frozen products whose primary characteristics were described in terms of nutrients rather than simply the foods that they were.[3] Combined with double hybridizing and now cloned processes, even before preparation there has been often massive technological transformation. Here, again, agribusiness and the food industry mirrors technoculture. In turn, this has provoked some philosophers to call for a revival of focal activities which include a sacralization of food preparation in the context of a specific region specifically, Albert Borgmann in his book *Technology and the Character of Contemporary Life*.

## 3.4 GLOBAL PROBLEMS

Most middle-sized problems belong, eventually, to larger sets. And as already intimated, modern technologies have made humans through it, global, geological forces. The first thorough awareness of this arose in this mid-century with the threat of global war, carried out with nuclear devices. The destruction of entire cities with single bombs, enacted with Hiroshima and Nagasaki at the end of World War II, became the symbols of this destructive power, enhanced many times over in magnificational power through both the development of the much more powerful hydrogen warheads and the superpowers' arms races which lasted until the very last decade of this century. While the *potential* for annihilation lies within any murderous or warlike intention, the actual capacity to carry it out as a species genocide occurred only with the rise of technoscience in warfare. The threat of nuclear, universal warfare, however, also cemented the consciousness of the interconnectedness of the *Earth as globe*.

The symbol for that phenomenon is also a result of contemporary hi-technoscience—the earthshot or view of Earth as a globe from beyond Earth. The drama of this blue planet as a globe is still a potent icon for late twentieth-century consciousness and is often used in various ways to depict global awareness.

Universal life destruction, a negative and extreme global possibility, remains today only a potential. With the collapse of the superpower contestation—and one could argue that one major source for this collapse was the rise of an equally technocultured consumer society which became irresistable in those areas of Europe deprived of its goods—the nuclear threat is at least momentarily backgrounded.

Instead, the more than actual Big Problem which forefronts the end of this century, is that of the global *environment*. In recent discussions the concern over the Greenhouse Effect with its associated ozone hole at the poles and general atmospheric pollution from acid rain to the rise in carbon dioxide levels, has captured the public imagination. Similarly, large scale industrial accidents such as

Chernobyl, Bhopal, and the Valdez oilspill incident, and the highly efficent experimental war, the Gulf War, in which disaster effects comparable to some earthquakes or volcanoes, also attain a high public visibility. In each case the late-twentieth century role of technopower as a geological-sized force is realized (negatively).

Such events give technological dystopians and determinists much satisfaction, and technological utopians little solace. However, one must also place such disasters and environmental degradation in a somewhat larger context: were one to take an up to date philosophical-anthropological survey of human cultures with respect to the variable of a positive environmental homeostasis, it would turn out that the picture prior to the twentieth century is, at best, highly ambiguous.

Many hunting and gathering societies were highly destructive of local environments, with only a minority attaining a genuine positive homeostasis. What masked this phenomenon was the combination of a very small population, often situated in a rapidly regenerative area which could recover from the destructive practices, such as rainforests. When the environment was limited and finite, this destructiveness became more obvious.

It is likely that late Ice Age humans were one major factor in the elimination and extinction of many large mammals which were among their favored game. Many of the Pacific Island cultures, as they spread eastward over much of the Pacific, clearly did hunt into extinction almost all wingless, and many other exotic birds—not only to serve protein needs, but often fashion ones. Exceptions to such often destructive practices can only be found in limited numbers of cases, such as those of the Inland Aboriginals of Australia, who for over 50,000 years had maintained a minimalist homeostasis with a desert environment. Or, the Arctic Inuit in a similar multimillenial balance with a sea and caribou-based food culture. Or, in some areas of pre-Modern southern Africa, the Central Plains of North America and other places were where practices which even included burning grasslands served to enrich the food chain of the largely human and animal scaped veldt.

I have already referred in passing to both the desertification which has frequently occurred with the introduction of domesticated animals and the degradation of a richer diverse Mediterranean area and its irreversible changes at the onset of our own civilizational sources. Again, isolated cases of positive homeostasis have also occurred within these same boundaries. European peasants learned to build up soils (after, of course, removing previous forest land) which sustained higher levels of population, and long centuries of flood cycle farming in the silt-rich Egyptian Delta, allowed for similar homeostatic continuities.[4] One can surely *not* make the case that environmental degradation is either a distinctly modern, or even a distinctly Euro-American phenomenon. One can only argue that the more massive and effective contemporary technologies speed up and raise the level at which such destructive practices may occur.

At most, one could argue that the currently possibly dominant tribes of technocultural societies have, indeed, let loose upon the globe high-powered and negative homeostatic forces. This is not to argue that modern technological impact, when negative, is identical with past similar circumstances. It is clearly obvious that the sheer magnificational power present in the ensemble of contemporary technologies far exceeds anything invented prior to this century. Moreover, as the plastics example illustrated previously, many of the products—including the whole range of esoteric toxins—are uniquely technologically constituted. To this, one might add that even raw materials, such as crude oil, have been concentrated into single large containers or conveyance modes which, when ruptured, cause larger effects. But all of these are made possible by the refinement of the technologies which allow more micro- and macromanipulations of entities previously either unknown, unconstructed, or unmanagable. These factors do differentiate modern technological capacities from traditional ones, but the cultural embedment into which such capacities fit retain older strands of societal values and possibilities.

All this is known to the modern and critically informed person. It should not be surprising, then, if a number of responses to the situation sketched have also arisen. The one position which today is endangered is the sheerly optimistic and utopian position which was more common in the nineteenth century and which held that what became technoscience, would in a short, foreseeable future, solve all of humankind's problems, introduce a leisure society, and create universal equality. Moving on a spectrum from more utopian to more dystopian contemporary positions, five of the many arguments taken in response to the concerns with the global problems are:

1) *Modern technologies are still developing,* and if left to open development, would evolve into more positive and refined technologies which could handle environmental problems. Just as the crude, large smokestack industries of the Industrial Revolution have begun to fade and be replaced by higher-tech processes and products, so will cleaner and more efficient technologies be developed to handle recognized problems. Of course to accomplish such ends, technoscience would also have to become more sensitive to its impact and implications and become more socially sensitive. Such a position might be called an optimistic refinement strategy. It implicitly argues that technoscience is basically correct in its premises and aims, but that even now it needs more development to attain its positive goals.

2) A second set of positions actually contains a variety of approaches, all of which are based upon a *scale-down* tactic. E. F. Schumacher's *Small is Beautiful* (1976) was a popular expression of such a strategy. Here the basic argument is that insofar as technoscience has become wedded to the macro-dimensions of Big Science, Big Technology, and Big Industry, control has become too difficult and centralized. A more decentralized and smaller scale approach is what is needed.

The appropriate technology movements of the 70s and 80s are also part of this strategy, here often combining a scale-down approach with a lowering of complexity to lo-tech versions of solutions. Often aimed at problems of technology transfer to Third World countries, appropriate technology proponents argued that the technologies ought to be small and simple enough to fit extant educational and cultural praxes.

But the same scale-down strategies can be seen to be worries even within the upper reaches of the technoscience echelons. The leading magazines concerned with science *(Science, Scientific American, Nature)* have all expressed concerns with the impact of maxi-funding of selected projects (Genome Project, Supercollider, Space Station) with respect to the continued support and funding for smaller and more innovative projects.

This family of positions takes the main variable for worry to be related to size (with its implied centralist and corporate administrative and funding structure). Its relations differ with respect to the degree of complexity and level of development, however.

3) More radical still is a group of *alternative technology positions*. This grouping is very diverse, but its major theses would be those which seek *alternative* models for technologies in both understanding and context. With respect to the environmental issues, deep ecologists, eco-feminists, and a whole new group of thinkers who wish to extend the range and scope of what had previously been thought of as human rights to non-human beings. One might call this movement a new rights movement, now including concerns for animals, even trees, and related to a primacy of environmental preservation.

The arguments here relate to the need to find different or *alternative metaphors* for the understanding of technoscience. Usually there is both a deconstructive and a reconstructive moment in the alternativist argument. As noted previously, many feminists see in the current situation, the long term result of a patriarchal, phallocentric dominance in technoscience. For technoscience to become genuinely beneficial it must both have its model of paternal dominance deconstructed, and, at the least, balanced with a more nurturing, caring or mothering substituted as an egalitarian alternative.

The new rights extenders, as I have called them, have a related but sympathetic argument. Here what must be deconstructed is the strong Western division between humans and the rest of nature, particularly animate nature, and the construction of a much lessened difference (if any) between humans and animate nature. While there is both a scientific and a religio-humanistic reason often given—much animal life is, at the genetic level, virtually identical to human life and thus respect of all living beings is to be established along rights' lines—the result is one which runs parallel to strands of more ancient animism or some strands of Eastern religions.

Basic to the alternativists, however, is an argument which directs itself at deconstructing/reconstructing the entire contextual understanding of technoscience. Sometimes the alternativists also wish to revive pre-scientific era contexts as supporting the alternative positions (Mother or Gaia Earth, animistic revivals, or variations upon these).

4) The most extreme position I shall call the *catastrophist* position. It is one which argues that modern technologies have developed to the point where they are no longer controllable and are run-away or automous. Technology is reified to a transcendental role, and insofar as it has become the dominant and alternative-destroying force of late modernity, has become virtually total in reign.

Here the argument is one which holds that the very success of technoscience is such that an irreversible trajectory has set in which, within the current epoch, is unstoppable. *Only a catastrophy of some sort could reverse the situation.* Dystopianism of this degree does not always provide an alternative view (no one can predict the result of a catastrophy) such as may be found among alternativists. Technology (with the "T") dooms humankind to either destruction, or some catastrophy which might (?) stimulate a re-starting of a different history.

5) There is, finally, a fifth position which does not belong neatly on the utopian/dystopian scale above. It is what I shall call the *postmodernist position,* elements of which may often be found in aspects of some of the above positions, but which in this context becomes a perspective which argues that *modernity* which includes the basic structures of technoscience, is already coming to an end. But what would replace modernity is not at all clear. Proponents of this position often are more optimistic insofar as the more dire projections of extreme dystopians are rejected, but, equally, some postmodernists become even more pessimistic insofar as they see a possible result of postmodernism as one in which a technological

virtual reality finally substitutes for all previous reality. Postmodernism may be taken as a position which takes *ambiguity* as the deep structure of technoculture. But it is also a position which rejects any return to past models as an outcome. Most postmodernist variants include affirmations of pluralist solutions, eclecticism, and a view which sees the Modern Era as coming to an end.

What all of the above positions acknowledge, is that there is a global, environmental crisis which must be addressed. It is the macro-problem of contemporary technoculture. But, even the conditions for addressing this crisis are not often met.

## 3.5 CONDITIONS FOR RESOLUTION

It is probably not at all unusual for human societal problems to outstrip the conditions for rational solutions. Indeed, in the contemporary world, such lack of conditions themselves become part of the overall debate. For example, it is clear that one widely agreed upon condition for rational argument about environmental crises would be to be able to accurately and quantitatively assess, *on a whole earth basis,* just what the state of the climate, the effects of acid rain, the ratio of forestation to deforestation, the ratio of desertification to new land production, etc., might be.

While much earth science is rapidly becoming more sophisticated with respect to such measurements, and while there is an emergent consensus of the scientific community on some of the variables, the last words are far from in. Moreover, even this basis as a condition can and does become part of the argument itself.

For example: is it apriori apparent that a sheerly quantitative analysis is either necessary or sufficient for resolving the issues? Some philosophers would argue that to give this much weight to a form of calculative reasoning is already to have skewed the debate into the arena of technoculture. Rather, the issues ought to be fought out on much more *qualitative* grounds having to do with the quality of life and the possible definitions of what a good life ought to be. They argue that to elevate sheerly quantitative considerations to the fore skews the debate in such a way that qualitative factors are backgrounded. Only when such qualities are isolated and affirmed does it make sense to judge what technoscience can and does do to enhance or detract from such life quality.

Or, even if one accepts a whole Earth quantitative analysis as one essential in the argument, its present insufficiency becomes itself a politically and ideologically charged issue. Conservatives argue that precisely because it is less that totally clear that acid rain causes x or y negative effects, one must delay action until all the results, or more of them are in. Contrarily, liberals argue that sufficient evidence is in to merit immediate social and political action which would change the practices of industry and government regarding technologically produced effects upon the environment. In one recent turn with respect to the Greenhouse Effect, while it is clear that the largest consensus is one which presumes that the warming cycle has set in, and that homogenic factors contribute an important element in this trend, there is still disagreement within the scientific community itself about when and how urgent action might be—must we start now? Or do we have about a century in which to act? All of this is grist for the study of the philosophy of technology.

But how? There is one sense in which, immersed in problems, the *phenomenon of technology* itself may slip to the background. Just how do humans experience technologies and how do these transform our very world? These questions must become the focus of the next section.

## CHAPTER FOUR

# TECHNOLOGY AS PHENOMENON

To this point I have traced two historical narratives: the first a tale of the interrelations of philosophy with science from its beginnings, with asides regarding its more recent relations to technology. And the second, an impressionistic look at technological developments in some of its high points, often related indirectly to, but also often non-coincidental to the development of philosophy.

I have also introduced representative contemporary problems now being discussed by philosophers of technology. As I have suggested, the disciplinary field of "philosophy of technology" is, however defined, relatively new. In an organized and institutional form it is no more than three decades old in North America. Like many new fields, it has been highly interdisciplinary and diverse, but today is beginning to be marked as a distinct sub-discipline precisely by the appearance of introductions, including this one. Textbook introductions often mark the end of a first phase, the formation of a sub-discipline, and the beginning of a second phase, a new academic field. In short, it is now time to introduce the very *contemporary field in the philosophy of technology.*

One of the tasks performed by an introduction, either implicitly or explicitly, is to present to the reader a discourse, the identification of a body of literature, and an identifiable scholarship. At the same time the field is already too large to discuss the full set of publications involved in its constitution. So, I shall profile only three approaches, but I prefer that the criteria choosing these be explicit and so I shall show my hand in outlining what I have used for inclusion and exclusion for the discussion to follow. Here is what I have considered in narrowing the field:

1) I shall focus upon recognizably *philosophical* works, although I shall take these broadly as a style of analysis which is rationally argued, critically developed, and comprehensive in scope. In the works to be noted, strong connections to some tradition of philosophy can also be noted. There are, of course, many very important works and authors whose backgrounds are in cognate fields which are relevant to the philosophy of technology. For example, I have not dealt with the long list of books by Lewis Mumford, one of the most prominent of intellectual historians, whose work has influenced and stimulated many of the philosophers today interested in technology (particularly his *Technics and Civilization,* 1934), nor with the often cited, but idiosyncratic Buckminster Fuller, whose praise of technology is often noted by others as an example of a celebrant of technology.

2) I shall deal only with works which take a perspective upon *contemporary technology as an overall phenomenon.* This criterion is to some extent arbitrary, since many excellent works have been more focused upon some particular problem set. For example, one of the most influential works on contemporary technology is Hubert Dreyfus's *What Computers Can't Do* (1972). And, as a technical study, Kristin Shrader-Frechette's *Risk Analysis and Scientific Method* (1985) is also important. Another more recent work, coming more out of comparative literary analysis, is Avatar Ronell's often cited, *The Telephone Book* (1989). And one could add a whole spectrum of interesting and related books.

3) And, again partly arbitrarily, I am restricting my analysis to books which may be regarded as both more comprehensive and possibly more mature works, rather than articles or multi-authored collections, in spite of their obvious importance in forming the field. Many of the earlier books—including the important series of *Research in Philosophy and Technology*—are thus set aside here.

4) Having already referred to the originators of a philosophical approach to technology (particularly Marx and Heidegger, but also

to the following generation of important thinkers, Ellul, Marcuse, Ortega, etc.) I shall concentrate here upon *living contemporaries.*

5) And, finally, as an indicator of a recognizable field of discourse, I shall narrow my focus to widely cited and *inter-cited* authors who would be thought to have produced the primary reference texts for the discussion. This, too, narrows the field, because a number of books in addition to those I shall deal with, have also appeared. And, I shall restrict myself to books in *English* for purposes of this North American introduction.

There is also a second, explicit way in which I shall preliminarily relate this introduction to the now growing field of introductions. I wish here to mention three earlier such introductions, relating each of them to national contexts, and take account of their cross-referencing. The earliest of these is Friedrich Rapp's *Analytical Philosophy of Technology* (translated into English in 1981, but first in German). Then, Jacques Goffi's *Que's Que C'est Philosophie de la Technique?* (1989), a prominent French introduction. And, finally, Frederick Ferré's *Philosophy of Technology* (1988), the first systematic American introduction to carry this title.

Among these nationally diverse introductions there is virtual unanimity about the originators and forbearers of the field. Karl Marx and Martin Heidegger both stand out as prominent in turning philosophy to the phenomenon of technology. Then, as earlier noted here, the field of critics including Jacques Ellul, Herbert Marcuse, and José Ortega y Gassett, are also noted.

There are, as might be expected, some differences as well. Germany has both the broadest and oldest traditions in the philosophy of technology, including the earliest books with *Technikphilosophie* as part of their titles (recall Kapp, 1877). Thus, Rapp's book includes the largest group of such German thinkers, including a contemporary of Heidegger, Friedrich Dessauer, *Philosophy der Technik* (1927), the same year as Heidegger's *Being and Time.* Unfortunately, Rapp was working too early to have been able to take account of the late 1970s/early 80s proliferation of contemporaries, particularly those in North America.

Jacques Goffi's introduction in France relates to a much more recent tradition and, again, cites many of the same thinkers noted in the field consensus. But he chooses four philosophers from different periods as important: Marx and Heidegger as the earlier formulators, then for the late-twentieth century, Gilbert Simondon, a French-speaking Canadian well known in Francophonic circles, and myself, Don Ihde, in the Anglophonic world as the two forefront contemporaries.

Frederick Ferré follows Goffi's penchant for a forefronted four, in this case Marx, Heidegger, Marcuse and—in praise of technology—Buckminster Fuller. He also includes Whitehead among the background figures, and for contemporaries includes Mario Bunge, Carl Mitcham, and my earlier works.

In addition to these three national introductions there has only very recently been a recognition that "philosophy of technology" is here to stay. Paul Durbin, the "godfather" of American philosophy of technology, has usually been reluctant to let the field be called that. Yet, in his introduction to *Broad and Narrow Interpretations of Philosophy of Technology* (1990) he in effect renames "Philosophy and Technology" philosophy of technology, and acknowledges Albert Borgmann, Carl Mitcham, Kristen Shrader-Frechette, Langdon Winner, and me as having the predominant international visibility in this new field.[1]

There is, then, both a core consensus regarding a small field of philosophers, and a set of differences often related to national contexts. For example, I have included here a highlighting of John Dewey as a primary figure in the background of the American context, to balance the similar highlighting of Dessauer (German) and Ellul (French) as forbearers noted in their respective contexts. However, in this chapter it is to the small group of living North American contemporaries who have published books which fit the previous criteria that I now turn. In this case I shall refer to three as follows:

1) Langdon Winner, a political theorist/philosopher, was the first to systematically approach the comprehensive notion of technology in the North American context with his 1977, *Autonomous*

*Technology: Technics-Out-of-Control As a Theme in Political Thought,* followed in 1986 with his widely read, *The Whale and the Reactor.* Winner was both the earliest and remains probably the most cited of the figures in this context.

2) Albert Borgmann, long a contributor to both philosophy of technology and one of its keenest analysts, entered the book field in 1984 with his *Technology and the Character of Contemporary Life.*

3) I followed Winner closely with the first North American book titled a "philosophy of technology" from a philosophy of science series, *Technics and Praxis: A Philosophy of Technology* (1979), then again with *Existential Technics* (1983) and *Technology and the Lifeworld; From Garden to Earth* (1990).

I do not pretend that the three authors noted here exhaust the field, although each does fit the criteria noted. There is a small group of authors who preceded our three, including Edward Ballard, *Man and Technology* (1978), Nicholas Rescher, *Unpopular Essays on Technological Progress* (1980), and James K. Feibleman, *Technology and Reality* (1982), and by press time for this book, a small flood of other books to be noted in the suggestive bibliography. But in terms of the deliberately narrowing and focusing criteria here, the three authors chosen are representative and suggestive for the state of the art.

## 4.1 LANGDON WINNER: TECHNOLOGIES AS FORMS OF LIFE

Of the three philosophers, Winner is the most widely cited, particularly within the social sciences. His background is primarily in political theory and philosophy and takes a unique social science approach to technology. His earlier book, *Autonomous Technology: Technics-Out-of-Control As a Theme in Political Thought* (1977) set the tone for much discussion to follow. And, as will be seen, his background figures include Marx and Ellul, but also Ludwig Wittgenstein, all previously noted.

Technology, as discussed in social science contexts, often would be fitted into what could be called a *technological deterministic* or a *social deterministic* category, with the latter the dominant one in most social science.

The social deterministic view is one which sees the development of technology arising largely out of power relations and the decisions of elites, or the groups of people in power. A classic, historical example of such an interpretation involves the early history of the McCormick reaper. In the nineteenth century, although there were several companies attempting to perfect a mechanical grain reaper, the McCormick Company took the forefront. During that development, the entire process of the casting of gears was done by skilled artisans who had mastered the entire process—but these skilled artisans also demanded monetary recognition of said skills, and threatened to form a union. McCormick, wanting both to retain the power of management and to break the nascent union, opted for a new process of gear manufacture which broke the process down into atomic steps, each of which could be undertaken by less skilled workers, *even though this process was slower and more expensive than the previous artisanal process.* This de-skilling tactic, associated with the name Frederick Taylor, was frequently followed in the early days of the Industrial Revolution by manager/capitalists as a labor controlling strategy.

Social determinists often chose such examples to illustrate how power elites *determine* technologies. However, by so focusing they frequently *implicitly also hold that technologies are, thus, neutral* and merely results of human decisions and manipulations.

In contrast, technological determinists hold that, once invented, technologies carry with their use a different kind of determinism which reflexively *forms society itself.* Karl Marx was, in a limited sense, a forerunner of such a view in his focus upon (technological) modes of production: "In acquiring new productive forces men change their mode of production; and in changing their mode of production, in changing the way of earning their living, they change all their social relations. The handmill gives you society with the feudal lord; the steam mill, society with the industrial capitalist."[2]

In *Autonomous Technology,* Winner took this theme—which by

the 70s had been elaborated by many of our "second generation" philosophers of technology: Marcuse, Ellul, Ortega, etc.—such that it had to be addressed by social scientists dealing with technology. It was this *theme* which Winner examined in his first book.

I shall not examine all the ramifications of the theme, other than to note that however hard or soft the technological determinist line might be, it clearly held that *technologies were not neutral.* And, this non-neutrality, when escalated to the complexity and interrelatedness of contemporary technology, carried with it more and more impact upon how society functions. Only at a certain stage could technology become autonomous.

*Autonomous Technology* took as its primary background figures Marx and Ellul. In a contemporary setting, Winner recites an interesting example of how a single technology—the snowmobile—followed a Marxian trajectory in reshaping the entire set of social and human/animal relations of the Lapps.

Prior to the introduction and use of the snowmobile, the social and animal interactive life of the Lapps had been one of a rather constant interaction between caribou and herders, such that the caribou could be considered to be domesticated and known virtually individually by the herders. Moreover, the process of care was one which was continuous throughout much of the year.

The snowmobile—which was enthusiastically adopted (as the history of technology frequently illustrates)—allowed the Lapps to move to the herds more quickly and comfortably than before. But in the wake of this single-technology adaptation, there followed a series of much larger consequences, among which were: (a) herding now occurred only at certain times of the year, (b) thus the human/caribou continuous relation was broken and the caribou became "wild," (c) but also the breeding cycle and calf care pattern was broken, and perhaps due to the regularly timed round-ups and the stress associated with this, caribou population began to drop drastically. And, (d) since fewer people were needed for the care of now smaller and more easily cared for herds, the caribou related populace of Lapps also had to change, with the reduction and breakup of the societal groups which previously were related to the caribou process. Here the Marxian insight that a different mode of production results in a different set of social relations is well illustrated.

And if in the context of a relatively isolated society, the introduction of a single technology had such large (non-neutral) ramifications, what of the overall implications in a larger, more technologically saturated society? Here the speculations of Ellul, who argued for a totalistic impact of technologies, appealed to Winner as well. Technology could become a culture, or—drawing from Wittgenstein—*a form of life*.

And although the Marxian and Ellulian themes do not disappear, the emphasis upon technologies as forms of life in a more technologically oriented form, became the focus of Winner's second book, *The Whale and the Reactor: A Search for Limits in an Age of High Technology* (1986).

The term, "forms of life," comes from Ludwig Wittgenstein, a philosopher of language often associated with Postivism in his early works, but transcending that position in his later work. Indeed, Wittgenstein's recognition that language games were complicated, interrelated human praxes far beyond the simplistic naming and reference orientations of philosophers of the time, gave a clearly *existential* cast to the notion of forms of life. Forms of life are *gestalts* which are often non-linear, but nevertheless recognizable. Winner applied these notions to technological development.

Arguing that too many (particularly social science analyses) of technologies simply looked at side effects and impacts, thereby often continuing the implicit belief that technologies are neutral, and employing a simple cause/effect mode of analysis, Winner argued that technologies create new worlds.

> We have already begun to notice another view of technological development, one that transcends the empirical and moral shortcomings of cause-and-effect models. It begins with the recognition that as technologies are being built and put to use, significant alterations in patterns of human activity and human institutions are already taking place. New worlds are being made. There is nothing 'secondary' about this phenomenon.[3]

Such technologies create forms of life. They are not neutral, but the way in which they come into being is also not simply causal at least not in any linear way. And while form of life is understood in a Wittgensteinian way, Winner continues to maintain much of the Ellulian perspective as well, a perspective which sees humans implicated in, taken into, technological *systems*.

The construction of a technical system that involves human beings as operating parts brings a reconstruction

> of social roles and relationships. Often this is a result of a new system's own operating requirements: it simply will not work unless human behavior changes to suit its form and process. Hence, the very act of using the kinds of machines, techniques, and systems available to us generates patterns of activities and expectations that soon become 'second nature.' We do indeed 'use' telephones, automobiles, electric lights, and computers in the conventional sense of picking them up and putting them down. But our world soon becomes one in which telephony, automobility, electric lighting, and computing are forms of life in the most powerful sense: life would scarcely be thinkable without them.[4]

Here, then, is a mode of analysis for technologies: what are the forms of life which emerge? How are they interrelated? What worlds are made through technologies? To make a technology is not simply to make a tool or an artifact—*it is to make a world.*

Winner's own focus remains social-political. In the second chapter of *The Whale and the Reactor* he announces the mode of analysis which he prefers in the suggestion that "artifacts have *politics.*" One of the examples he works out was in my own Long Island neighborhood—the bridge system on the Long Island parkways planned and developed by the notorious archplanner, Robert Moses.

The parkway system on Long Island is, indeed, aesthetically pleasing. The four lane highway weaves along treed avenues and under a system of uniquely different stone bridges, about which there have been photo essays. But the bridges are not neutral in a subtle, but at first effective way. They are low enough to clear automobiles, but too low for either trucks or buses, and this was deliberate: Moses was building a politics into the bridge system, one designed to keep the (lower class) public out of the Long Island suburbs. They did not often have cars, and, to help the matter along, Moses vetoed any extension of the Long Island Railway to even such popular places as Jones Beach. Once in place, the bridges provided a long lasting and clearly expensive to change politics, favoring exactly the upper middle classes who fled New York City to Long Island. The bridges belong to a wider social praxis in their non-neutrality.

Winner detects at least two levels at which artifacts embody politics: the first is one in which technologies provide convenient means of establishing patterns of power, although with a certain flexibility within the pattern set. A second level is one in which a technology is more closely linked to more particular political and social forms. For example, nuclear power—if for no other reason that the highly dangerous weaponry possibilities of its byproducts such as plutonium, and the toxicity of spent uranium—*must* be carefully (and therefore, centrally) controlled.

It is the growth and interconnection of particularly the latter sort of technologies, with the inclination of the operators working these technologies toward often undemocratic patterns of action, which worries Winner and stimulates the questions of limits for high technology.

The closing symbol of the "whale and the reactor" poignantly outlines this worry. Returning to his boyhood home area on the coast of California, Winner finds the much disputed Diablo Canyon nuclear reactor:

> Below us, nestled on the shores of a tiny cove was the gigantic nuclear reactor, still under construction, a huge brown rectangular block and two white domes. In tandem the domes looked slightly obscene, like breasts protruding from some oversized goddess who had been carefully buried in the sand by the scurrying bulldozers. A string of electric cables suspended from high-energy towers ran downhill, awaiting their eventual connection to the power plant.... [then, noting the surf and Diablo or 'Devil's Rock'].... At precisely that moment another sight caught my eye. On a line with the reactor and Diablo Rock but much farther out to sea, a California gray whale suddenly swam to the surface, shot a tall stream of

vapor from its blow hole into the air, and then disappeared beneath the waves. An overpowering silence descended over me.[5]

And Winner uses this experience to condense his own experience of high, twentieth-century technology, and to wonder if there are *limits.*

His conclusion is that contemporary high technology does display a pattern, not unlike that mentioned earlier in the discussion concerning Heidegger's about technology as taking the world as standing reserve or a resource well:

> If there is a distinctive path that modern technological change has followed, it is that *technology goes where it has never been.* Technological development proceeds steadily from what it has already transformed and used up toward that which is still untouched.[6]

It uses up what has gone before and proceeds to that which is yet unused. And that is why the question of limits becomes crucial for Winner. Technologies are not only forms of life, they are, in the contemporary hi-tech sense, totally expanding forms of life. Winner, although more cautiously than his predecessors, joins again the worries expressed by our "second generation" philosophers of technology.

## 4.2 ALBERT BORGMANN: THE DEVICE PARADIGM AND FOCAL THINGS

Although Borgmann's *Technology and the Character of Contemporary Life* (1984) does not fit the chronological progression of this parade of authors and books of the "third generation" of philosophers of technology I am discussing, his position on technology does fit here.

Winner's background figures were Marx, Ellul, and Wittgenstein. Borgmann draws from Heidegger, and not unlike Heidegger himself, also from Aristotle, (but this is the Aristotle of the *Ethics,* not the Aristotle of the *Metaphysics).* And there are both similarities and differences to be noted in comparison with Winner.

Like Winner, although from a different tradition, Borgmann would clearly hold that technologies are like forms of life—they belong to complicated and non-neutral human praxes. But, like Heidegger, Borgmann sharply distinguishes *modern* technologies from their older traditional roots. But while the modernity which characterizes high technologies is usually associated with the *scientific* character of contemporary technology, in Borgmann's case it is the progressivism and implicit optimistic (capitalist) *liberalism* of modern technology which is focused upon.

Modern technology in this ideological context, carries with it an implicit *promise:*

> Technology...promises to bring the forces of nature and culture under control, to liberate us from misery and toil, and to enrich our lives. To speak of technology making promises suggests a substantive view of technology and is misleading. But the parlance is convenient and can always be reconstructed to mean that implied in the technological mode of taking up with the world there is a promise that this approach to reality will, by way of the domination of nature, yield liberation and enrichment.[7]

Such a promise, held in some respects since Francis Bacon, but heightened in the Enlightenment, is the motor which culturally drives modern technology.

According to Borgmann, however, the results are quite different from the promise. The promise seduces the modern *toward* a focus upon material goods, commodities, and a deep kind of quantitative thinking which, in turn, leaves the ancient Classical (Greek) and Christian questions of a *good (or excellent)* life out of the equation.

How technology does this may be seen in the Borgmannian symptamology which emerges in *Technology and the Character of Contemporary Life* between the device paradigm and focal things. The contrast between two kinds of "things" echoes much of Heidegger.

One of Borgmann's favored focal things is the *hearth.* In its technological history, it could even have been seen as a kind of technology. A hundred years ago in Montana (Borgmann's chosen region), the warmth of the hearth could not be taken for granted.

Indeed, most of life was unheated—coaches, the outdoors, sleighs—and thus when homes were finally built and fireplaces or other forms of hearths were installed it could have been seen as a distinct improvement.

But a hearth, once established, can be a focal thing: it gathers into a centered place the activity of the settler, farmer, rancher, or homemaker a whole set of praxes. These include the processes of cutting the trees, splitting and drying the wood, building the fire, cooking and warming the house, etc. In this way the hearth belongs to a kind of world, "A thing, in the sense in which I want to use the word here, is inseparable from its context, namely, its world, and from our commerce with the thing and its world, namely engagement."[8]

How the hearth thus gathers a set of *focal activities* is illustrated thusly:

> A stove used to furnish more than mere warmth. It was a *focus,* a hearth, a place that gathered the work and leisure of a family and gave the house a center. Its coldness marked the morning, and the spreading of its warmth the beginning of the day. It assigned to the different family members tasks that defined their place in the household. The mother built the fire, the children kept the firebox filled, and the father cut the firewood. It provided for the entire family a regular and bodily engagement with the rhythm of the seasons that was woven together of the threat of cold and the solace of warmth.[9]

In principle, one could see in this analysis a form of life, the making of a world, and a certain mode of production *à la* Marx, Wittgenstein, but particularly of Heidegger and his artful mode of *technē,* with the hearth as a kind of technology/object of art.

The good, the excellent life, which emerges from Borgmann is one in which there is home, family, the enjoyment of seasons, regional cuisines and home cooking, running, trout fishing, in short, many of the virtues which could be associated with his preferred Montana lifestyle.

It would be too easy to caricature this lifestyle as a contrast to what has emerged from modern high technology—for there is here a certain romanticism which also echoes Heidegger's preference for the Black Forest, hand tools, stone bridges over steel ones, and the like. But it is also a romanticism which is widely and deeply shared by many North Americans.

If a hearth is a focal thing, what is a device? Borgmann argues that at the heart of modern technology lies a *device paradigm*. It embodies the promise of technology to disburden its citizens from what is burdensome. As a counterpart to the hearth, then, central heating plants may be seen as devices. The world of central heating is radically different from the wood-burning hearth:

A device such as a central heating plant procures mere warmth and disburdens us of all the other elements. These are taken over by the machinery of the device. The machinery makes no demands on our skill, strength or attention, and it is less demanding the less it makes its presence felt. In the progress of technology, the machinery of a device has therefore a tendency to become concealed or to shrink. Of all the physical properties of a device, those alone are crucial and prominent which constitute the commodity that the device procures. Informally speaking, the commodity of the device is 'what a device is there for.' In the case of a central heating plant it is warmth, with a telephone it is communication, a car provides transportation, frozen food makes up a meal, a stereo set furnishes music.[10]

In Borgmann's view, the device is a type of technology which commodities by means of a certain functional abstraction—mere warmth, communication, transportation, a meal—and which conceals certain aspects of an effortful human set of praxes which belong to its counterpart, the focal thing. The device *reduces*, by means of its concealed machinery, everything to a means-ends function.

At this point one could see that both Winner and Borgmann hold that (a) technologies are clearly not neutral; (b) they generate patterns of human praxis or worlds; and (c) that modern technologies have, in effect, taken over larger and larger territories of that human praxis. But here, now a certain divergence emerges between our first two authors.

Winner is worried over what might be called technological totalization, with respect to *limits* and to a certain centralist and non-democratic control often associated with high technologies. Borgmann, although clearly agreeing that the expansion of modern technology holds the field, in effect, argues that modern technology is associated with a kind of democratic liberalism, but that this very liberalism is what has failed. Modern technology has failed to live up to its very promises in spite of delivering much of the liberation from toil, disease, and the harshness of pre-Modern life: "The promise of technology was one of liberty and prosperity. But the brilliance and joy of life that are implied in the promise have not come about in spite of two centuries of gigantic efforts."[11]

The failed promise, however, is not total. Borgmann grants that much of the promise *has* been fulfilled, but in the very fulfilling of that which we may take for granted, there lies a potential hollowness:

> The technological means that have freed us from hunger, disease, and illiteracy have become part of the inconspicuous periphery of everyday life. The commodities that fill the center of our lives with entertainment and diversion gratify us in a passing and shallow way. We take justified pride in the intricacy and power of the technological machinery that we have constructed and continue to improve. But this confidence about the means goes hand in hand with great diffidence about the ends in which they issue.[12]

In short, once the calamitous threats of starvation, early death, and ignorance are conquered, does "life [become] ruled at its center by triviality and frivolity?"[13]

If this is the dominant way in which modern technology has formed a world, it is not, however, the only alternative. Borgmann argues that technology is *reformable*. It is reformable through the recognition of, revival of, and enhancement of *focal activities*. I shall look briefly at two of his examples.

Two of Borgmann's pet whipping boys are frozen food and "canned" music. Contemporary food processing can easily and simply provide the home with quick food. Here is a device which produces a commodity. It conceals the way in which food would be prepared in a focal form, which at the most extreme might entail the householder to have a garden, to pick the fresh food, and go through the entire process of preparing the meal. Borgmann points out that modern technology does not eliminate this possibility—indeed, there are frequent signs that precisely such attention to quality food selection and preparation is often enjoyed and undertaken precisely in a civilization which has enough leisure and freedom from pre-modern worries such that a meal can be an aesthetic, even sacramental event.

Similarly, while reproduced music via the stereo is a virtual commonplace, it is also in some sense a commodity for consumption. It is merely enjoyed, not made. Borgmann celebrates, again, the focal activity of *making music*. What could be called home production is another way in which the intimate group—family and neighbors—can be gathered around a quasi-sacramental practice. And, again, the wider world of a technological culture which has secured life against hunger, disease, and ignorance, does not preclude such focal activities as such. They can, as it were, occur in the interstices of a high technology culture.

Borgmann sees signs that interest in such focal activities frequently occur. They are forces against shallowness and superficiality—and even if rare, those who participate in them come to see a richness which he hopes will flavor technological life.

Revivals of the joys of full meal preparation, the almost cult-like activity associated with running, craft developments, and other focal activities can and do, although in limited cases, serve positive purpose and are not contradictory to technological culture: "If our lives are centered in a focal concern, technology uniquely opens up the depth and extent of the world and allows us to be genuine world citizens. It frees us from the accidental limits of shortness of time, lack of equipment, or weakness of health so that we can turn to the great things of the world in their own right."[14]

## 4.3 DON IHDE: LIFEWORLD TECHNOLOGIES

Chronologically, my *Technics and Praxis* (1979) follows closely on the heels of Winner, and *Existential Technics* (1983) and *Technology and the Lifeworld* (1990) sandwich both Winner and Borgmann's works. The background for all three books lies in phenomenology in the broad sense, and draws from Edmund Husserl, Maurice Merleau-Ponty, and, of course, Martin Heidegger.

*Technics and Praxis* began with a set of observations about the way in which science (and philosophy) is related to technology and praxis and argues that one should reinterpret much of the modern science in the light of its technologies, particularly *instruments.* I undertook an extensive, but preliminary phenomenology of instrumentation to show a variety of *human-technology relations.*

Human-technology relations, patterned after a phenomenological analysis of human intentionality, purport to show what is invariable in the ways humans experience their technologies. For example, *embodiment relations* are uses of technologies which enhance (and non-neutrally *transform)* our perceptual-bodily experience of an environment or world.

In the case of the sciences, the early use of optical technologies, such as telescopes and microscopes, revealed worlds heretofore not expected. But the very magnificational powers of early optics also oriented inquiry towards the *macro- and microworlds* revealed. As such, the instrument transformed not only what was seen, but its scale in relation to non-instrumental vision.

What emerged from the analysis as a structural feature of instrumental use, was what I called a magnification/reduction transformation. For every enhancement of some feature, perhaps never before seen, there is also a reduction of other features. To magnify some observed object, optically, is to bring it forth from a background into a foreground and make it present to the observer, but it is also to reduce the former field in which it fit, and—due to foreshortening—to reduce visual depth and background. Such non-neutral transformations belong to all technologies.

If embodiment relations enchance (and reduce) bodily- perceptual experience, *hermeneutic* (interpretive) relations take another mode of reference to observed objects. Here the analogue is to reading and language rather than sensory perception, and is exemplified in instrumentation which uses various forms of measurement (dials which use numbers or spectra, etc.). The object is still being referred to, but is now translated into a dial reading which indicates some more abstract (and thus more reduced) aspect of the object, such as weight or heat. And the process requires a special reading skill which knows how the instrument refers.

Both such human-technology relations exemplify ways in which humans—with technologies or instruments in a mediating position—experience an environment or world in a new or technological way. But such activities do not exhaust human-technology relations as others are more of a background character. For example, automatic or semi-automatic machinery—such as Borgmann's example of central heating—may function in the background and not occupy any focal attention. One may be experiencing the heat, but barely if at all aware of the switching which is going on and off (unless the system breaks down). Here technological systems begin to function as quasi-environments or technological cocoons within which our daily lives play out.

It can easily be seen from this early set of examples that many of the features of technology in my analysis correspond to similar emphases in Winner and Borgmann. Like them, I was arguing that technologies are non-neutral in the human context function like forms of life or worlds, although in different ways with different technologies, transform human experience.

Nor does the transformation of human experience stop with the directness of sensory or first person experience. In *Existential Technics* (1983) I turned to some of the reflexive ways in which a growing technologically mediated experience of the world reflected back upon such phenomena as human *self-interpretation* and its cultural variants.

I argued that it was not at all accidental that the primary metaphors for explaining bodily functions should be technological ones—hearts are "pumps," brains have "wiring," language learning is "pre-programmed," etc. Rather, this is a reflection of a basic and immediate enviromental texture which, for the late modern, is *a technologically textured one.*

While both the above works were, in some sense, preliminary, *Technology and the Lifeworld* (1990) much more systematically outlined the theory of the technological lifeworld which I see. Like Winner and Borgmann, my approach has been one which takes patterned praxes as basic. Such patterns form gestalts which change from human historical period to period, and also from *different human cultures.* But there is both a structure and a variant upon structure to the human experience of technology, I argued.

Human-technology relations—such as those which implicate our bodily-perceptual activities—are structurally crosscultural. And in *Technology and the Lifeworld,* I drew from both many historical and different anthropological contexts to show how this was the case. But at the same time, technologies in the ensemble are also *culturally embedded.*

One of my favorite examples was drawn from the history of navigation. Western navigation, at least by the time of Columbus on, but perhaps even earlier, was instrument-mediated (one found one's position by means of astrolabes, speed by log, etc.), mathematically interpreted (the globe was a map divided by ideal lines of latitude and longitude), and perspective privileged as overhead (a reading position).

In contrast, South Pacific navigators—who successfully explored and populated virtually all of the inhabitable Pacific islands by a millenium ago—did not use instruments, privileged their own bodily position in the sailing craft in a relativistic set of directional interpretations, and read the stars, wave patterns, bird paths, etc., as the mode of orientation. The two modes were both successful, but were gestalted very differently and with different technologies (contrast the vessel technologies of mono- with multi-hull boats).

To this point one might see much in common with the analyses of Winner and Borgmann, although the perspective of *Technology and the Lifeworld* is much more multicultural than the more standard Western orientations of their works. However, when I turned to the strictly contemporary issues also discussed by Winner and Borgmann, a certain set of differences emerged.

Both of our previous authors hold that modern technology is now a world phenomenon, and I agree. Both take it that such technology "goes where it has not been" or moves toward a kind of totalization, and I again agree. But, I argue, the totalization is *presumptive* and at this juncture is beginning to show signs of serious strain which may harbor quite different directions.

Modern technology and technoscience is clearly an invention originating in Western culture. It has clearly "englobed" the Earth. But that is, while dominant, only one outward and expanding moment. I argue, with a metaphor of a tide with an undercurrent, that the undercurrent is one in which increasingly the underside of the dominant is the growth of two closely interrelated phenomena: (a) the first is the non-avoidable awareness of *Others,* i.e., non-Western cultures. This awareness is part of the communications technologies, particularly the image technologies (such as television, cinema, and all forms of visual networking) which daily bring us exotic cultures and makes clear the conflicts between cultures. This very presence makes what had heretofore been able to be an unspoken cultural heritage itself become a matter of necessary choice. My culture is different—hence, in some degree *arbitrary,* and thus it cannot automatically be taken for granted. But, (b) secondly, this multicultural undercurrent is itself multiple. In our image technologies, it is fragmented into culture bits which, in turn, become part of the now *postmodern* awareness. It appears as a kind of cultural *bricollage* such as MTV also shows in its image form.

I then argue that what is distinctive about the emergence of a postmodern moment is a different kind of vision—a *plurivision,* which is symbolized by a kind of insect-like, compound vision. Our world is one of multiple screens, like television newsrooms, which carries news from many sectors, but from which we must edit and mix to create a coherent, but also multiple-sourced program which, ultimately, becomes our collective autobiography.

The summary metaphor for postmodernity and what it may, but not necessarily can, imply is a culinary one. Today's cosmopolitan world is a culinary eclectic. On Mondays it might be grandmother's meat and potatoes; on Tuesday, Hunan Chinese; on Wednesday, Northern Italian; on Thursday, French *nouvelle* cuisine (already Orientalized); on Friday, Mexican spicy, etc. Even, occasionally, "fast food." Here, then, is a conclusion which *celebrates* a certain disappearance of a "core," or a "foundation," and which is both anti-romantic and anti-nostaligic, modes of perspective which I often detect in the second generation of philosophy of technology (Marcuse, Ellul, etc.).

However, this is not to say that this divergence from the set of worries exemplified by Winner and Borgmann are absent here, they merely are taking different form and direction. Our biggest worries, I am arguing, ought to be *global,* first in the sense of concern for the Earth's environment, and second, in finding post-enlightenment means of securing intercultural (and thus also interpolitical and intersocial) modes of tolerance and cultural pluralism. The first entails *limits* as Winner emphasizes, and the second a new species of intercultural agreements which also must *limit* the cultural-religious forms of negative totalization which today characterize many global conflicts.

## 4.4 STATE OF THE ART

Although I have chosen to highlight only three philosophers, I do not claim to have exhausted the contemporary field of philosophy of technology. It will be seen, however, that there are both wide areas of agreement, but also very different tonalities to the analyses just undertaken.

Regarding some of the most important areas of agreement, I have underlined which might be called a world or gestalt analysis of technology which is a theme of all three of our authors. Negatively, this is to say that *philosophically* all three reject a simple means-ends or neutral tool analysis of technologies. Technologies are contextual and belong in different ways to *praxical gestalts.* They are also multidimensional with respect to their role within human experience and culture. Analyses which restrict such a larger perspective run the danger of *concealing* the full impact of any technology.

Nor can one philosophically be restricted to some simple set of objective classifications of technologies as to type. This is particularly the case with respect to what I have sometimes called the designer fallacy. Only sometimes are technologies actually used (only) for the purposes and the specified ways for which they were designed. Two interesting examples of this have been the typewriter and the telephone.

Both were originally intended as helps for impaired persons, the typewriter as a possible way for blind persons to write, and the telephone to assist the deaf or hearing impaired. What was to become their extremely important set of social uses ultimately entailed little of the original designer intent.

In the case of the typewriter, as it began to be adapted for both copying and composition, there followed a massive reorientation of the secretariat. Before the typewriter, most secretaries were male; but Luddite-like, many males rejected this new keyboard mode of writing. Women, culturally accustomed to the keyboard (musical), quickly adapted to this new mode of writing and soon dominated the secretariat. Other changes, of course, entailed a whole reorganization of such things as business offices.

All our philosophers also agree that technologies are non-neutral, although each focuses upon different aspects of the transformational powers of technologies in use (Winner on socio-political dimensions, Borgmann on social and ethical values, Ihde on perceptual-cultural dimensions). Again, negatively, this is to reject a sheerly instrumental interpretation of technologies. Just as designer intent gets transformed in actual use, so do mere expressions of purpose get changed in the adaptation of technologies. This is what makes the question of technological control so difficult.

The question is never so much a matter of controlling technologies, since even the simplest technology (such as a dip ink pen) has an ombra of counter control on the user. I am used as much as I use any technology, whether on a first person or a social level. As Winner shows, technologies have politics, or as I show, they have cultures. To control technologies, particularly in the ensemble, is much more like controlling a political system or a culture then controlling a simple instrument or tool, again, particularly in a contemporary high technology setting.

And, all our authors agree that the contemporary setting is one in which the complexity and extent of technologies is unprecedented, with equivalently unprecedented degrees and types of human, social, and cultural transformations. But when it comes to both worries and recommendations, nuanced differences occur (Winner focusing upon limits, Borgmann upon reform via focal activities, and Ihde upon dealing with technopluralism).

The philosophy of technology, clearly, is not itself technical, nor should it be. It must be a philosophical reflection and critique of the phenomenon *technology.* Its role, as in any "philosophy of..." is not unlike that of literary (or other) criticism to a body of specialized activity. But also, like any new critical field, toleration for, let alone appreciation of, criticism is not always immediate or easily taken. But, no more than writers, or artists, or performers (or, in this case, technologists) can get rid of their critics, however friendly or antagonistic, so the newly established field and traditions of philosophy of technology will persist despite their critics.

What is important is that the critique which a philosophy of technology implies be one which respects the complexity of the phenomenon. The suggestion here is that the selective field of philosophers displayed here, by taking the phenomenon technology to be more than some collection of artifacts, more than means-ends tools, point out some of the culture.

Technologies reveal worlds or world-aspects; they are non-neutral in the ways in which the human-technology uses emerge, and that at several levels (perceptual, socio-political, cultural); and they display—particularly in the ensemble—a variety of implications for human histories.

The current state of philosophy of technology, noted here in its "third generation" of philosophers, has just begun to undertake this work. Thus the question of not only what problems are important contemporarily (some of which were noted in chapter three) but of the future also arises. It is to that set of issues that I next turn.

CHAPTER FIVE

# FUTURES

Technologies are now forefront phenomena in our daily lives. And the philosophy of technology has begun to critically reflect upon these phenomena. Where do we go from here? This is to imaginatively speculate upon near futures, both of technology and the philosophy of technology. I shall here look at certain trends and return, in part, to some of the previously noted problems, now to be looked at in a different context.

We have seen that any issue implicating contemporary technologies is multidimensional and complex, and that it takes on a certain gestalt characteristic. In what follows I shall be ranking various larger problems to suggest directions of the future of philosophy of technology, and by way of using imaginative variations try to show some of the complexity which each problem set shows.

## 5.1 THE ENVIRONMENT AS FOUNDATIONAL ISSUE

I would today argue—and I suspect there would be considerable consensus on this—that the issues relating to the global environment are the most important for philosophy of technology to address. That is because there is a basic foundational issue related to the environment. Were the environment to become so degraded, or irreversibly damaged, that humans would find it hard or more difficult to live in, its foundational status would become obvious.

Yet, as strange as it may seem to many, such an urgent concern with the environment has not attained the political, social, or cultural importance which it would need to become *the* forefront problem for the beginning of the twenty-first century. Let us take an analogue for another such foundational issue: human health—*our health*—is another phenomenon which carries this weight even without our explicitly saying so. And, because it is an *unspoken foundational issue* in our culture, it affects many other issues.

For example, medical doctors constitute the highest paid profession in North America. The mean income of an M.D. is higher than that of the mean for a lawyer—and much higher than that of either private or public financed scientists, (and much, much higher than the mean for mere philosophers!). In my university, the highest paid scientist is a Nobel Laureate physicist. His salary, while very respectable, comes to less than the mean for a practicing M.D. and is less than half that of the salary of the dean of our medical school. One reason why the public tolerates this economic hierarchy, in spite of the fact that medical knowledge is certainly no more technical, complex, or nuanced than that of several of the other professions mentioned, is that the bottom line is that *without basic health, all our other activities become either more difficult or, even, impossible.* Health is perceived as a foundational concern.

Were the environment so perceived—particularly on a social level—it might be the case that the activities associated with its preservation or restoration, too, would rise to the top of our willingness to pay scale. Yet, it clearly *is* such a foundational issue. But the very diffuseness and widespread nature of its associated problems, and the lack of an as yet sense of emergency has been insufficient to cause a reshuffling of priorities and attention upon it. As a society we are not yet ready to tolerate the high cost for environmental care that our medical establishment has for personal health.

Once in a while there do occur signs which begin to worry the public. Technologically implicated disasters, the Valdez oil spill, Bhopal, the Challenger explosion, Three Mile Island and Chernobyl, and the eco-terrorism of the Gulf War, have all frightened many. But in spite of these shocks to contemporary awareness, the tenacity with which North Americans hold to the desire for cheap energy (particularly automobiles and gasoline), certain kinds of environmentally degrading employment (any industry engaged in destruction of nonrenewable resources), and a basic perceived standard of living which is environmentally wasteful, outweighs any dramatic change in direction.

There is, of course, a public awareness that the environment is an issue and there has been no dearth of positions relating to how the problems may be solved. We have noted several of these previously, but here I shall return to only three main strategies to simplify the issues. One might be thought to be anti-modern technology; one a compromise situation; and one a pro-Modern technology position.

(a) There have been a variety of suggestions revolving around a pull the plug strategy, i.e., recommendations to give up high and complex technological developments. There are many variations upon such a strategy, but they usually contain the elements of *i.* returning to some previous lower level of technology or to minimalist technologies entirely, *ii.* some nostalgic element which believes that a previous time (or other culture) had it better and simpler, and *iii.* often an implication that it is our culture which is most guilty of the current ills.

What if one were to drastically and immediately pull the plug? Apart from the social disruption, which always would produce unexpected results, several other probably unacceptable results would also have to take place. For one, world population levels would have to decline drastically and quickly. This would have to happen if any decent overall standard of living were to be maintained. Any opting for prior levels of even balanced human/environment strategies—such as those of the Inland Aboriginals in Australia, the Inuit in Alaska, slash-and-burn cultures limited to small areas in rapidly recoverable rain forests, or, for that matter, nineteenth century, more rural and agricultural societies—all entail very much lower population levels with respect to any given area of Earth.

Such a move—imaginatively equivalent to a population decline comparable only to times of vast warfare or plagues—would clearly entail also persuading a population of the need not just for lower levels of self-reproduction, but actual decreases in population self-replacement levels. This would be a major cultural transformation, for which there is *only one extant model to date for willing compliance*. And that model turns out to be ironic for the pull the plug position holders. *Only those societies which have attained a high standard of living in a technologically saturated context have willingly and without external constraint limited population growth to self-replication or lower.* Most European countries, North America, and, more recently, Japan, best instantiate such a population leveling or lowering family size set of habits.

The countries which have tried to officially limit births through political programs (for example China and India) in the light of clearly rational perceptions of the impact of population growth upon standard of living have not notably succeeded.

The picture can also be complicated further—even with lowered family size, say from five to ten children per family down to even three or four—modern sanitation and medical practice have significantly lowered the infant mortality and mean (not horizontal) longevity of virtually all populaces. Surely, the pull the plug strategist would not suggest that modern medicine also be given up in favor of less life preserving health strategies. Persons arguing for alternative medical models, other than technologically impacted ones, usually argue for their equal or even greater efficacity in preserving life.

As simplistic as it is, any genuine variant upon a pull the plug strategy would entail results as catastrophic in their own way as much of a predicted environmental catastrophe.

(b) A second set of strategies, less drastic, but clearly directed against what are perceived to be both the centralist and most complex of contemporary directions, might be called *ameliorationist* strategies. These strategies accept certain hi-tech procedures and developments, but turn them into regional and decentralized modes of such items as food or energy production.

For example, in the 1970s there developed a group of imaginative planners on Cape Cod called the New Alchemists. This group of strategists focused much contemporary environmental knowledge upon what could be called multidimensional but small scale energy and food intensive production. Small scale algae producing tanks, warmed by solar energy, fed fish raised in the tanks. Solar and wind produced energy drove the electricity needs of the experimental plant, etc.

On a larger scale, much contemporary experimentation with aquaculture—from catfish raising ponds in the southern United States, to salmon farms in New Zealand—have sought to supplement, and ultimately perhaps replace, much ocean or stream fishing which previously supplied the same food products. Today's oceans, undergoing overfishing with new hi-tech means, may well be analogues of the buffalo prairies of the nineteenth century in America. It may be noted that the buffalo—once replaced by domestic cattle and raised in much more limited and controlled situations—is no longer needed so far as food production is concerned. I am not arguing here in any way that the senseless killing of the buffalo by so many colonialist settlers was a good thing. I am arguing that the use of natural resources such as fishing the oceans may have very bad depletion effects upon species, and that for the preservation of species we *may need to find substitutes* such as are potential in aquaculture. One could go on with the food/population problem, to which must be added the major factor which threatens our environment, climate change, and atmospheric and oceanic pollution, all now clearly affected by homogenic—human (technologically implicated)—causes.

Ameliorationist strategies differ from the more extreme pull the plug strategies in that they concentrate contemporary knowledge into decentralist, but hi-tech solutions.

Yet, the rate of technological proliferation and change outpaces most alternative strategies as such. Such changes seem to outstrip efforts to ameliorate development or reduce such strategies to sectarian-like alternatives in the face of an established religion.

(c) A third strategy might be called hi-tech reformist. It affirms the current direction of hi-tech development, but argues that such directions as miniaturization, clean-up technologies, and alternate energy production (and some here argue for more rather than less nuclear produced energy) may provide the answer. This strategy also affirms the necessity of a more centralist, management-model approach. For example, a *world* agency to monitor and police energy consumption would be entailed.

The Dutch put forth a plan some years ago, a carbon credit proposal. All countries would, based upon resources and population, be assigned a carbon credit limit. Third World countries could trade such credits to the technologically advanced countries in exchange for technological education and development programs, etc., in the hope that an overall world solution to slowing down Greenhouse Effect gas emissions might buy time for all until even higher-tech solutions might be found. This strategy assumes both a continuation of the current and even enhanced centralist planning mechanisms and an evolution of ever higher-tech developments, but concentrated upon clean solutions in contrast to the present "dirty" ones.

There are many other variants, but in the cases just cited each strategy does recognize that the quality of the environment is a *foundational* issue and that it must be addressed. But this issue is also enmeshed in a complex *cultural philosophical* set of issues which entail nests of interrelated concepts. For example, the concepts technology, environment, and nature are all interrelated.

For instance, our very concept of environment is related to our notion of, and attitude towards nature, which in turn, entails a concept of what it means to be human, and into all of which must be fitted our means of interaction with each of these dimensions, technology. At this level we are at a very high altitude of *philosophy.*

Were we to take only one such dimension—nature—it would quickly be seen that today there is no consensus on what counts as nature. Again, I shall look at three popular versions of the concept of nature.

(a) Heidegger, as noted earlier, argued that the Western metaphysical traditions tended to take Nature as a kind of resource well *(Bestand,* "standing reserve") available for human use. Here one strand of the Biblical tradition taken in a certain way—humans as the images of God for whom nature is there to dominate—and the

Greek materialist tradition—again nature as mere matter—combined with Baconianism—nature to be changed through human knowledge-power—lies behind this concept of nature. This view of nature, in various forms, might even be taken as a dominant concept within much Euro-American technoculture.

(b) That concept is today challenged by variants upon a revival of interest in an older animist set of traditions which see nature as itself lifelike. Earth is our mother or alive, and to be not only respected, but quasi-worshipped. Recent attempts to elevate animal life into a quasi or equal (human) status belong, in part, to this revival. It appears among many animal rights groups, in some variants of eco-feminism, and in deep ecology movements. It is frequently tinged with strands of Eastern religious thought. It is also sometimes associated with genderization by some, who would seek a return to a more feminized version of nature. All of these variants, insofar as they have occurred as minoritarian views within Euro-American domains, are reactions against the previous dominant position.

Although I have lumped together and oversimplified both the nature-as-resource and the nature-as-animate positions, one can see that the directions involving technologies will be quite different for the two extreme positions.

(c) Today there is also emerging something of a hybrid concept in the development of Gaia or Gaia-like variants, where the whole Earth is seen as a kind of biological/geological system which, while not necessarily holding that the Earth must be all alive as in animism, functions as an organically self-correcting system. Although its originator viewed it as a way of conceiving the Earth as a single living being, it can also be seen to be an organic/inorganic interactive system.

It could quickly be seen that technologies would take very different shapes and roles in each of these variants upon the meaning of nature. In the first, at least in its most extreme forms, if technologies are the ways in with power/knowledge is embodied, the more powerful the technology the more likely to be able to change or dominate nature the better. This is often identified with the contemporary Western traditions of high, industrial technological development which, today, is also often thought to be masculinist or paternal in orientation. Here one finds contemporary technoscience which arose within and is associated with this alternative.

In the animist-related version, technologies would not be absent, although they might be more minimalist. Here the emphasis might well be upon cottage industry, non-corporate agriculture, craft- and home-produced products, etc. The earlier emphasis on village and cottage production favored for India by Mahatma Gandhi might be a good example of a modern thinker who responded to technology in this way. Most small is beautiful and appropriate technologies approaches could also correspond to the implied attitude toward technologies implicit here.

What might not be fully appreciated is that to enact this alternative, not only must there be a genuine revolution in the culturally dominant concepts, but that both what counts as science and technology must also change. Technoscience arose within and remains associated with a certain victory of, first, a largely *mechanical* set of metaphors. And although these have been challenged and have become much more subtle, sometimes replaced by electronic and even linguistic metaphors, a similar metaphorical power for the animist alterative has not appeared. For that reason (and others) it remains questionable whether there could *be* an animist science (although there clearly could be an animist technology).

The hybrid Gaia organic systems approach is possibly ambiguous with respect to its possible technologies. Do, and can technologies play a role within the self-correcting interaction of bio/geological systems? One could so argue, so long as such technologies could amplify or enhance corrective directions rather than disruptive ones. This view entails a vision of human activity which is out of balance with the organic rhythms and their timing within a whole Earth. Thus, at the least, technologies, whether high or low, would have to be brought into the delicate balance which the organic/inorganic Earth processes display.

I have been quite speculatively offering these ideas to show not only how essential the environmental issue is as a foundational issue, but how interwoven are the complex gestalts which surround any consideration for environmental correction or preservation. But at this level, while technological gestalts can more clearly be seen to be very different depending upon which notion of nature prevails, we are also very far from being able to deal with particular technologies or their development. We are at too high an altitude to be able to relate either particular strategies or particular technologies to the holistic nature-environment notions just surveyed.

Yet, within the matrix of solutions, even at this level, one can detect some areas of technology/ideology interaction. In what follows with each of the directions I shall examine, I shall at least look at one technology or technological set to indicate something of the subtle role it plays even with respect to the alternatives. Let me speculatively predict just one such interaction. I predict that the Gaia hypothesis approach to environmental analysis and problems will rapidly gain ground in the twenty-first century—but not for the usual reasons which might be given.

The usual way in which theories are thought to gain ground have to do with combinations of evidence and the satisfaction which the theory provides in accounting for the phenomenon which it purports to deal with. I am not dismissing these factors: increasing evidence of bio/geological interactions is occurring. Adjustment of $CO^2$ levels related to the plankton populations of the ocean have become increasingly important in analyzing directions of the warming phenomenon. This factor, along with many others, shows an increasing interaction between global biological factors and the climate (although it is far from proving that the Earth is or should be considered a single, organic unit).

My prediction concerning the probability of an increase in popularity of this model of explanation, however, relates to a different factor altogether, a *technological factor.* In this case it has to do with the increasing use of a certain instrument in Earth-related sciences: the computer modelling process. The Gaia hypothesis is a systems hypothesis in which there are a very large number of interacting variables—it might be called a complex, finite system—*which is precisely the kind of system which is both challenging, but manageable, for computer modelling.* One can plug in changes in one or many variables and project what changes will occur in the computer generated model. Moreover, the results are both useful and dramatic for educational purposes, and, with graphics, are perceptually graspable. A program called SimEarth is already popular in use for Earth sciences classes and choses a variant upon the Gaia hypothesis for its model. I am thus suggesting that the use of the computer, while not determining a direction, inclines our very inquiry in that direction. The instruments we use affect our results, and in this case the more complex the phenomenon, the better the computer model at present is for that interface.

This is also to suggest that whatever changes of directions are taken, they will be taken in the context of an already developed set of technologies, which, in turn, relate to how possibilities are seen and undertaken. In the just noted set of options, insofar as they entail overall views of things, I am suggesting that the very instrumentation which we use to analyze our phenomena, also provides us with *trajectories of inclination.*

## 5.2 PLURICULTURE

If there would be broad consensus over the most foundational issue facing philosophy of technology—the environmental crisis—there would be considerably less agreement concerning the next direction which I shall outline. That direction is the emergence into increasing prominence of the phenomenon I call pluriculture.

It would be generally agreed that the emergence and development of contemporary high technology is a largely modern and culturally Euro-American phenomenon, which today has been particularly successfully adapted into several Pacific Rim Asian cultures (Japan foremost, but also into Korea, and some of Southeast Asia). I cite only those countries—and I would include in this list those technologically European-derived locations such as Australasia, Israel, and South Africa, which are also already developed—which have sufficient technoscientific infrastructure so as to be both relatively autonomous and scientifically innovative.

Modern technology, of course, is also a world phenomenon since there is no country without relation to that development. However, after the above named locations, most other countries and cultures remain at the level of taking in and adapting to high technologies. These include the largest countries, population-wise, China and India, which while approaching some degree of autonomous infrastructure have not yet so arrived.

This is to say that with respect to development and production, only a limited number of countries and cultures are today modern with respect to technologies. But with respect to *reception* of technologies, all countries have been effected. And, at the forefront of that effect are the *media,* or what I shall call image technologies. Image technologies are, at this popular communications level, something of a counterpart to the just noted use of computer-determined systems analysis. That is, image technologies are the dominant modes of contemporary communications and provide the focus for inclined trajectories. Print, radio, cinema, and television are, effectively, universal. By this juncture at the end of the twentieth century, the major symbols of technology transfer are no longer the steel axes and plows of the nineteenth century, but radios and televisions. But, also, such media as *communications* are always two-directional or interactive.

It may be said that it is the technologies which do the interlinking, and insofar as it is this system of machinery, the dominant outflow has been from Euro-American sources toward all others. But, with respect to what is carried by these technologies and embodied through these technologies, one can say that image technologies have also brought to universal awareness a multiple awareness of Others. This is an intercultural phenomenon closely tied to communications technologies and constitutes a kind of undercurrent to the outward dominant current of technological diffusion.

Its importance to the future of technological civilization, however, has not yet been fully appreciated. I have suggested or hinted earlier in the text that historically and with some frequency, periods of technological innovation have often coincided with multicultural periods. The early hunger for spices in Europe brought in its wake new ideas for power machinery; the voyages of discovery in the fifteenth and following centuries revolutionized much of everything from European cuisine—think of potatoes, tomatoes, corn, etc.—and materials—and earlier, I noted the correlations of technological development in the equally culturally diverse Hellenic-Roman period.

I have, secondly, noted that technologies are frequently variably culturally embedded. The same technology may be used very differently and even mean very different things in two different cultural contexts. Daniel Boorstin's parallel history of the mechanical clock in eleventh century China is one in which all time is isolated in the Imperial domain in contrast to the European early public use of time; this is related to the calculation of the Emperor's progeny and their horoscopes, in contrast to the European regulation of work time; etc.

Might we not expect there to be, in the early twenty-first century, just such proliferation and adaptation on an intercultural basis? The answer, I believe, is positive and may be illustrated in the following observations: automobile production, particularly in its most technologically advanced form, was once primarily a Euro-American phenomenon. Today it is the Japanese who are increasingly the primary producers and developers of the car. What is not so often noted is that they have also introduced a culturally different style of precisely that development.

When in Spain in 1989, late at night after the sessions of a philosophy of technology conference, I chose to relax by watching on television a major European automobile rally being held in hilly, muddy conditions in northern England. The racers were finely tuned stock models—and in the end all but one of the top ten winners was of *Japanese manufacture*. One reason for this success was that the winning cars had four wheel drives and most four wheel steering, which in those conditions simply ran away from the European models which lacked these features.

I cite this example, first, to note that the myth of mere copy technology which pervades much Euro-American belief about Japanese technological traditions is not only false, but that we miss what is a very different cultural motor driving this development. The Japanese cultural mode is one of both high hierarchy, but also of group consensus. Such a form is, undoubtedly, slow for development if contrasted to the American, in particular, emphasis upon the individual breakthrough—presumably someone having a brilliant idea and bringing it into existence. What was missed, however, was that the very consensual practice of consistent, if small, improvements which also belongs to Japanese car manufacture, had, ultimately, a dramatic accumulative result. Few Europeans or Americans now miss this effect with respect to product quality, and they are beginning to see its results with respect to the gradual development of innovation itself. Again, the same technology is differently embedded with respect to cultural praxis, but it is no longer clear that *only* Euro-American praxis is the root source.

From a different perspective, in the recent past much explanation of both change and development was given a primarily *economic* cast—and, particularly with respect to the vast Third World/First World contrasts, one cannot dismiss this—there are signs that conflicts, changes, etc., should also be regarded from a crosscultural perspective. The collapse of Eastern Europe communist governments in the 80s/90s decade juncture, while clearly related to economic factors, was also due to media enhanced *perceptions* of a vast difference between consumer standards of living between eastern and western Europe. But in that case the basic cultures were already family resemblant.

Nevertheless, some of the same intercultural features may be found between Euro-American and other less-related cultures: much of Asia has become highly consumerist and in successful adaptations. But much of Africa and South America, in spite of the social desire to become so, has not so attained the same level of affluence and commodity-rich levels. And, in spite of the virtually worldwide appearance of a desire for high technology commodity culture, the form resistance and reservations take is also informative of the overall appearance of pluriculturality I am pointing to.

One of the most interesting test cases regarding cultural change relates to the role of women in the various cultures undergoing modernization. Here we meet again the nest of issues which relates to population size and family planning as a crucial variable. Human populations have always practiced some forms of population control and birth control, but the forms have been widely variant and often, for modern sensibilities, have been quite horrendous.

In ancient times exposure (practiced in certain Mediterrean societies, for example), infanticide (practiced widely in the South Pacific), and, particularly, the killing of female infants at birth, were several of such practices. Modern birth control technologies ranging from the Medieval invention (or, possibly earlier) of the condom, to both the Pill and barrier methods of the twentieth century, have taken the edge off many termination strategies by replacing them with non-pregnancy strategies. And, combined as we have seen with industrially produced levels of affluence, these have led virtually all of the high infrastructure societies to a distinctly modern family size, rarely larger than self-replicative size. But the very acceptance of such family strategies has so far failed to occur in precisely those parts of the world where the combination of economic affluence, cultural resistance, and non-technological infrastructure levels have not occurred.

I suggest that much of the reason for this lies in the complex set of social roles which are more cultural than economic as such. Remaining for the moment with gender role variables, another feature of many of the countries where birth rates have significantly declined, one may also note degrees of modification to the roles of women with respect to the private and public sectors. In most industrialized countries, women have increasingly taken up roles in the public domain. In Russia and the United States both careers and family roles for women are common. In societies in which there is industrialization, but which also retain a much higher degree of association between women and the private sector—for example, Japan—the current model is one of temporal periods. Many women, if not most, in such societies gain education and, for a time, take up careers, but after which they return to what was earlier the more traditional home role. Yet, even the periodization of public/private is a changed variable with respect to gender role.

A particularly complex example of both a conflict and appreciation of gender role occurs in contemporary Islamic countries. Some feminists have argued that Islamic culture has a gender role pattern which is satisfying to many women, in that the clearly strong division of male/female roles in those societies is such that there is a strong and relatively autonomous sector and community dominated by women (and a second sector similarly dominated by men). Yet, at the same time, there are clearly signs of great impatience and frustration over cultural restrictions and religious limitation effected particularly in the more modernized Islamic countries.

The examples above, however, focus primarily upon the impact of technological modernization upon traditional forms of culture. I have elsewhere argued—and I continue to hold this position—that contemporary technoculture is *acidic* to all forms of traditional (mono)cultures. Thus I wish to turn briefly to another facet of pluriculture.

Image technologies everywhere potentially or actually make present a variety of Others—including an amazing variety of "exotic" cultures. World news of world political events are good examples of this phenomenon. One cannot not be aware of the cultural clash both within the Middle East between the related, but also very different Semitic cultures, but also between Islamic and Western cultures. In a less politicized domain, world travel, television, movies, and other forms of crosscultural exchange, have made more and more persons aware of Otherness. But this awareness is not without its own form of technocultural effect.

To illustrate this, I shall revert to my favored culinary metaphor: I seriously doubt that what could be called a core cuisine exists or can be taken for granted anymore in any cosmopolitan setting—although it can exist regionally and be there both deeply embedded and made into a version of Borgmann's focal activity. But in New York, or any other major world city, ethnic food in its fullest variety is the order of the day; Asian: Korean, Vietnamese, Chinese (in at least four regional varieties), every European variety, Mexican, the eclectics such as *nouvelle* cuisine which combines French and Oriental themes, etc., only begin to suggest the varying list.

The connoisseur is the sampler who soon learns that there *is not a clearly superior cuisine* even though some may be generically better than others; there may well be healthier/less healthy, expensive/less expensive, or other considerations which do and should enter the considerations. He or she learns to celebrate the variety itself—a different cuisine for a different day and occasion. I insist on celebrating the Chinese New Year just as strongly as I continue to celebrate the traditional North American Thanksgiving.

The pluripalate, however, is a metaphor for something much deeper. It is a metaphor for a unique kind of contemporary multiculturality which is increasing its impact in technoculture. In many areas we can increasingly pick and chose, almost *bricolage* fashion, among the various items of exotic cultures and blend them into a contemporary, eclectic melange. This includes not only cuisines, but fashions, literature, philosophy, and even religions. And only a superficial look at this phenomenon would regard it as a surface phenomenon.

Its deeper implications are much more serious: (a) pluriculture implies that contemporaneity is no longer simply Euro-American as a dominant form. Were it that, the more specific values of Eurocentrism would be identified. Rather, there is here a kind of both/and set of values—often very relativistic, but clearly multiple. Such a value set has often been termed, *postmodern.* (b) The very awareness of competing and multiple cultures makes our own traditions much more a matter of deliberate choice—if we are to insist upon maintaning them—and thus a certain cultural *arbitrariness* is also implied which already demythologizes the taken for granted quality of *a* cultural tradition. And (c) the emergent form of a *bricolage* picking and chosing is itself distinctively different from many past and traditional cultural forms, and is perceived as such particularly by those which see in this phenomenon its potential threat to traditional forms. (d) Finally, I would argue that contemporary pluriculturality is also not Western, but neither is it Eastern, or any other traditional form. It is, rather, beginning to show its own distinctive characteristics which are often simultaneously international as well as preservative of certain regional or parochial-ethnic shapes.

In short, I am arguing that pluriculturality is *already an emergent and modified form of culture which is no longer simply or distinctly Euro-American.* And, as I have noted, its emergence is closely associated with the prevalence of the image technologies which have—like all technologies—begun to change social practice and form.

## 5.3 WARS AND WEALTH

Insofar as there have been ensembles of technologies associated with the first two our our future directions, it may be inferred that the ensemble most closely associated with the environmental crisis is that of our industrial and consumptive-related technologies. I have not argued, and do not hold, that *all* industrial and commodity-related technologies have negative impact. But, accumulatively, it is clear that with respect to levels of pollution and Greenhouse Effect enhancement, there is an overall negative and dangerous effect.

Contrarily, with respect to the emergence of pluriculturality, the single largest group of technologies was seen to be communication (image) and transportation (travel and cultural exchange) technologies. I am well aware that not all persons would regard the multicultural cosmopolitan result as positive—many, detecting the acidic effects upon traditional forms of culture, regard its emergence as negative, threatening, and superficial. Yet, insofar as pluriculture does follow the culinary metaphor, its effects are to have sharpened our appreciation of Others and to have dulled the worst possibilities of cultural chauvinism.

It may also be seen now, that the environmental crisis is primarily focused upon the human/enviroment or Earth interface, while pluriculture focuses upon the intercultural, intellectual, religious, and aesthetic sensibilities of humans. There remains a set of problems which must focus upon *social, economic, and political* issues, particularly upon the problems of *justice* thereby implicated. I have thus titled this dimension of the contemporary, wars and wealth.

Very much of precisely this industrialized First World technological development has been *military.* North Americans, particularly those in the United States could hardly escape this fact in the last decade or more. Technoscience itself has been deeply implicated in this process and every science discipline is in some degree dependent for its research money upon defense or military-related projects (ranging from 35 percent support for mathematical sciences up to 85 percent for aeronautical sciences).

Military technologies, of course, have a long history and were we to take World War II as a turning point, one can see much has changed in the five decades since that war. As I wrote this book, I was in Australia during the Gulf War which dominated the news. I cite the following experience and observation: last night I saw the movie, *Memphis Belle,* the historical tale of a B-17 bomber's last mission over Germany in 1943. At that time the Allied Air Forces were losing up to a quarter of their bombers on each raid. Moreover, newspapers comparing the Gulf War and World War II bombing results indicated that in World War II it took an average of some nine thousand bombs to eliminate a single, major target (a factory, for example). Were we to switch to the (clearly controlled) evening news, wherein it was emphasized that we employ "smart bombs" which are laser guided to go (not unlike the movie, *Star Wars*) into ventilator shafts and explode the target from the inside out. And while the intimation is that it now takes one bomb to do what nine thousand did fifty years ago, even if one doubts that difference, the ratio is nevertheless a vastly different one due to the technologies involved.

On a grander scale, one may note also a difference in the role and ratio of civilian to military death and injury in the five decades involved. By the end of World War II it had become very difficult to separate military from civilian casualties—the fire bombing of Dresden, followed by the atomic destruction of Hiroshima and Nagasaki in no way distinguished between civilian and military targets.

The subsequent Cold War armaments race, through the development of nuclear bombs hundreds of times more powerful than the "little boy" of Hiroshima, and multiplied into thousands of warheads, brought the superpowers to a stand-off with respect to nonconventional weapons. The probable result of an international nuclear exchange had become the technologically amplified intent to kill escalated to species genocide! All winners thus also became in the same exchange losers.

Instead of large scale international wars, micro-wars entailing terrorism, and regional wars often entailing First World/Third World countries (the latter aided by their First World natural resource users) took their place. The Gulf War reintroduced a form of warfare which theoretically could again differentiate between military and civilian sectors in a hi-tech version of long past warfare. Ironically, insofar as that becomes possible, it again makes war "thinkable," and hence very dangerous for contemporary life. That was part of its experimental nature. This is not to say that civilian suffering did not take place, but in some senses its character was changed. It became indirect in terms of the side-effects of the Gulf War through accidents, through malnutrition, through health threats in the destruction of infrastructures, etc.

This is also to note that *precision*—which is clearly one of the developmental trajectories of much high technology—is as essentially ambiguous as a structure and feature of contemporary technology as all its other features. Precision—for example, in laser technologies—is possible as a surgical tool as in cataract surgery, as a measuring means in geographical measurements between the Earth and the moon, and as a laser guided "smart bomb" as above.

Contemporary warfare now belongs very intertwined with our first direction, issues facing the environment, as well as the second, cultures, particularly in their potentially negatively conflictual modes. If there has been a (temporary?) collapse of economic and political form conflict (liberal democracies plus capitalism vs. communist monoparty forms plus centralist socialism), no less virulent has been the emergence of cultural-religious conflict (Jews vs. Arabs, Islam vs. the West, ethnic strife in regional conflicts, Protestant vs. Catholic, Hindu vs. Moslem, etc.). Contemporary warfare thus encompasses both our first two directions—enviromental and cultural ones.

Wars, however, also are implicated in a much larger social justice issue—the disparity of wealth distribution on the Earth. Here social and political philosophy, read via technologies, also can come into play. The most macro-features of First World/Third World disparities have to do with the flow of natural resources from the latter to the former, while the power relations are usually attempts to control from the former to the latter. The Third World perceives the latent power embodied in the hi-tech military, communications, and consumerist forms of the technological society and usually—despite reservations and even resistance to what is equally perceived as a threat to traditional cultures—seeks to gain a hold onto these technologies. Many of the most tragic consequences of the last century have had to do with which technologies get distrubuted first, which second, and which third.

The disparities are often enough pictured: the *National Geographic* magazine some years ago printed a photo from India, in which a farmer was pictured in the foreground plowing with a wooden plow with only a steel tip, drawn by an ox, against a background in which a television satellite was pictured, pointing upwards to its geostationary space relay. The image, however, also reflects a national policy which is attempting a leapfrog strategy, which through communications is attempting to bring sectors of Indian life directly into the beginning of the twenty-first century.

More shocking are the images we have all seen of mine-maimed Cambodian peasants, still attempting to live a traditional tribal-village life, but victims of often superpower support for a political group which presumably supports more or less this or that superpower's interests.

Technology transfer, philosophy of technology, must, like all other technological phenomena, be seen as always more than mere hardware transfer. I would argue that much of the recent decade or so past appropriate technology movement often simply missed the point insofar as it focused upon the transfer of technologies which were presumably more indigineously easily introduced and gradualist. If Winner is right concerning "artifactual politics," the very non-neturality of all technologies and particularly of technology transfer, then the insertion of what, when, and how technology transfer is to occur is one of the most crucial issues for the socio-political dimensions of philosophy of technology.

## 5.4 PHILOSOPHY AND DEVELOPMENT

In this introduction I have tried to survey much of what has formed philosophy of technology, and have gone from low to high altitudes. Much of what I have implicitly suggested is that philosophy of technology attempts a reading of contemporary phenomena in the light of our praxes and technologies. It is, in this way, a *perspective* upon contemporary problems.

I wish to conclude, however, with another suggestion with respect to the role philosophy (and philosophers) might take with respect to technological development. I shall develop two actual examples here, with respect to this issue: philosophers and medical ethics, and philosophers and artificial intelligence. Respectively, the first example draws upon a long traditional concern of philosophers in ethics, and the second in an equally long concern for epistemology.

One of the earliest institutional involvements of philosophers with new technologies and their implications, occurred in the development of medical ethics sometimes also thought of as a special branch of applied ethics. At first, it was not even apparent that any "philosophy of technology" was involved.

I have already referred to this development, noting that through such institutions as the Hastings Institute, and in the incorporation of philosophy as a subject matter and philosophers as part of a medical education faculty, medical ethics became part of both much medical education and of health care. And I wish to affirm that I think these developments are positive and necessary and that not only in the medical context, but in other applied ethics contexts such as in business schools or in engineering schools, similar approaches should be encouraged.

But at the same time, there is a certain anomoly which may be detected in such a use of philosophy/philosophers when read via a philosophy of technology perspective. There is a sense in which much extant use of philosophy/philosophers in applied ethics contexts comes *too late*. It is too late in the sense that the very place at which the activity of philosophers occurs *presupposes* that all the technologies are in place, the development having occurred, and the problems which result must now be fixed up. Philosophers are to apply a philosophical or ethical band-aid to effects which have resulted from new medical procedures and technologies.

This is not to argue against uses of band-aids which sometimes must be used. And, clever philosophers can and do have a way of getting at things which are much more basic than providing technological fix ethical solutions to contemporary dilemmas. I leave that to them. But it is to suggest that the place and position for genuinely helping change is at a much more basic level—it is at the level of *development itself, particularly of technological development*. Here few philosophers dare to tread.

One place in which some philosophical treading did occur, with interesting and significant results, is in what I shall call the Dreyfus/artificial intelligence debates which began in the 1960s and only now are culminating in new developmental directions.

Hubert Dreyfus was a phenomenologist teaching at MIT in the 60s. He developed a critical interest, epistemologically, in the then current artificial intelligence, and cognitive simulation development then going on. He was hired by the Rand Corporation to investigate why the extravagant claims made in the 50s and early 60s concerning breakthroughs in computer artificial intelligence did not seem to by paying off as predicted. The result, first a study titled *Alchemy and Artificial Intelligence,* (1967) and later a book, *What Computers Can't Do* (1972), was a scathing critique of the metaphysical and epistemological models which motivated and lay behind the artificial intelligence models of the day.

He contrasted what he took to be gestalt, informal, and open-ended human thought processes with the atomistic, formal, and closed-system processes entailed in computer software models. At first the computer people reacted strongly and negatively and dismissed the Dreyfus critique, and many programs were attempted precisely to refute Dreyfus, only some of which succeeded and thereby indirectly improved various programs.

Many, however, continued to fail and the Dreyfus/artificial intelligence debates continued for more than a decade, *until a new generation of designers began to take his critique—and his implicit alternative models—seriously.* Those who did take his notions concerning praxis, bodily motion, and gestalt patterning seriously, began to develop different strategies of design. For example, one rather directly involved with Dreyfus was developed by Terry Winograd and called ontological design, and thus of a different technological direction. I cite this instance, not because the debate is either over or a single set of victors emergent, but because it illustrates a very different position for the role of philosopher vis-à-vis technology development. It is a foundational position insofar as the philosophy/technology interface here is one at the developmental level. I am thus suggesting that in a contemporary technoscience environment, one important future direction for philosophy of technology ought to be aimed at the research and development level and not only at the already developed and status quo level.

That is more likely to occur only if and when philosophy of technology increasingly finds its own place in distinctive junctures within educational and research-oriented institutions and, concretely, when both action and published results become more widely known in those contexts.

The third incident is autobiographical and relates to some of my earlier studies of technology and perception. In 1989 I traveled to Japan for my first visit and, again, to present a paper at a conference. There I met a former design engineer who had heard a paper I was giving on human-technology relations in a Pittsburgh area context in the early 70s. The paper had been on embodiment and hermeneutic human-technology relations and entailed a detailed discussion concerning perception and instruments. At the time he was employed by a consulting firm for airports and their safety designs.

He indicated that in the previous decades a number of crashes were attributed to misreading of approach signals and lights. The perceptualist approach I was taking brought the *Geistesblick* or "Aha" phenomenon to his thinking and he indicated that he went out and "stole" the approach, justification, and explanation I had used in writing up a proposal for the redesign of approach hardware. I had been focusing upon the way perception formed gestalts and was mediated through technologies. The account was highly relevant to the perceptual errors made by pilots, but which were inherent in the approach designs. His firm won the contract and the approaches were redesigned with the desired effect. His comment was, "You probably didn't know that you were [indirectly] responsible for saving many lives." This is, however, another example of philosophy at the foundational or developmental level. It is where much philosophy of technology ought to occur.

I am not suggesting here that philosophers have any special design expertise, or that an informed national policy would necessarily place them on all think tanks relating to research and development, but I am suggesting that the way in which a philosophical analysis, ideally relating the complex of ideas and dimensions pointed to in this introduction, needs to be an element in our developmental thinking.

Philosophy and philosophers have come late to technology. But in the contemporary world, with technology as a forefront phenomenon, it is timely that critical thinking, some of which has been most finely honed in the centuries of this oldest academic profession, be related to that phenomenon. Critical thinking, of course, is far from limited to the examples just cited. Rather, the whole spectrum of concerns from epistemology to ethics, from socio-political philosophy to aesthetics, could bear upon relations to technology.

This introduction is thus one set of suggestions about what has been and could be done once one's attentions are seriously directed to contemporary technoculture.

# NOTES

## CHAPTER 1.

1. Mario Bunge, "Five Buds of Techno-Philosophy," *Technology in Society,* 1 (Spring, 1979), p. 68.
2. L. Sprague DeCamp, *The Ancient Engineers* (New York: Dorset Press, 1963), p. 93.
3. Lynn White, Jr., is perhaps one of the best known historians of technology of the Middle Ages. His *Medieval Technology and Social Change* (Oxford: Oxford University Press, 1962) is a classic in the field.
4. The term, "technoscience" may have originally been coined by Gaston Bachelard in the 30s, but it has come to popularity in recent times primarily through Bruno Latour's *Science in Action* (Cambridge: Harvard University Press, 1987).
5. Roger Bacon quoted in Carlo Pedretti, *Leonardo da Vinci's Machines* (Firenzi: Becocci Editore, 1988), p. 12.
6. An excellent exposition of Bacon as experimental theorist is to be found in Ian Hacking's *Representing and Intervening* (Cambridge: Cambridge University Press, 1983). The Bacon quotations are from the J. Robertson (ed.) *The Philosophical Works of Francis Bacon* (London and New York, 1905) and are quoted in Hacking, p. 246.
7. Ibid., p. 246.
8. Ibid., pp. 168–169.
9. Jacques Ellul, "The Technological Order," in L. Hickman, *Philosophy, Technology and Human Affairs* (College Station, TX: Ibis Press, 1985), p. 40.
10. Ibid., p. 40.
11. Ibid., p. 45.
12. Ibid., p. 45.
13. Ibid., p. 46.
14. Herbert Marcuse, *One Dimensional Man* (Boston: Beacon Books, 1968), p. 7.

15. Ibid., p. 7.
16. Ibid., p. 19.
17. Jürgen Habermas, *Toward a Rational Society* (Boston: Beacon Press, 1970), p. 98.
18. Ibid., p. 104.
19. Ibid., p. 107.
20. Ibid., p. 111.
21. Alfred North Whitehead, *Science and the Modern World* (New York: Harper and Row, 1972), p. 107.
22. Martin Heidegger, "The Question Concerning Technology," *Basic Writings* (New York, Harper and Row, 1977), pp. 303–305.
23. Larry Hickman's recent book, *John Dewey's Pragmatic Technology* (Bloomington: Indiana University Press, 1990), has placed Dewey centrally in the philosophy of technology discussion. It reclaimed what had previously been taken as Dewey's instrumentalism and correctly relocated Dewey as a protagonist in the traditions being discussed here.
24. Dewey quoted in Hickman, op. cit., p. 291.
25. John Dewey, *The Later Works* (Carbondale, IL: Southern Illinois University Press, 1989), 12:162.
26. Ibid., 12:434–435.
27. Dewey quoted in Hickman, op. cit., p. 292, my italics added.
28. John Dewey, *The Middle Works* (Carbondale, IL: Southern Illinois Press, 1989), 8:78.

## CHAPTER 2.

1. See Daniel Boorstin, *The Discovers: A History of Man's Search to Know His World and Himself* (New York: Vintage Books, 1983), for many interesting discussions of technologies in multiple cultural contexts.

## CHAPTER 3.

1. Sandra Harding, *The Science Question in Feminism* (Ithaca, NY: Cornell University Press, 1986), p. 114.
2. For a full development of this unusual consensus between Anglo-American and Euro-American philosophers of science, see my *Instrumental Realism: The Interface Between Philosophy of Science and Philosophy of Technology* (Bloomington: Indiana University Press, 1991).
3. Lawrence Busch, "Science, Technology, Agriculture, and Everyday Life," *Research in Rural Sociology and Development,* vol. 1, pp. 289–314.
4. Rene Dubos, *The Wooing of the Earth* (New York: Scribners, 1972 has a par-

ticularly strong set of examples of various positive human-environmental balances within European and other histories.

## CHAPTER 4.

1. Paul Durbin, editor, *Broad and Narrow Interpretations of Philosophy of Technology* (Dordrecht: Kluwer Academic Publishers, 1990). See especially Durbin's introduction to the field, pp. ix–xvii.
2. Karl Marx, *The Poverty of Philosophy* (New York: International Publishers, 1963), p. 109.
3. Langdon Winner, *The Whale and the Reactor* (Chicago: University of Chicago Press, 1986), p. 11.
4. Ibid., p. 11.
5. Ibid., pp. 165–66.
6. Ibid., p. 174.
7. Albert Borgmann, *Technology and the Character of Contemporary Life* (Chicago: University of Chicago Press, 1984), p. 41.
8. Ibid., p. 41.
9. Ibid., pp. 41–42.
10. Ibid., p. 42.
11. Ibid., p. 246.
12. Ibid., p. 247.
13. Ibid., p. 248.
14. Ibid., p. 248.

# SUGGESTED READINGS

Although references within the chapters may be used as suggestions for supplementary readings, particularly with respect to classical authors, I am here listing a number of twentieth-century sources which may be used to expand upon the narrative of the text.

### I. THE HISTORY OF TECHNOLOGY

(I have used illustrated and easy-to-read books, often related to television series on this topic.)

Daniel Boorstin. *The Discoverers*. New York: Vintage Books, 1985.

James Burke. *Connections*. Boston: Little, Brown & Co., 1978.

L. Sprague de Camp. *The Ancient Engineers*. New York: Dorset Press, 1963.

John Merson, *Roads to Zanadu*. Netley: The Griffin Press, 1989.

Sigvard Strandh. *The History of the Machine*. New York: Dorset Press, 1979.

### II. FORERUNNERS OF THE PHILOSOPHY OF TECHNOLOGY

Jacques Ellul. *The Technological Society.* New York: Alfred A. Knopf, 1964.

Jürgen Habermas. *Toward a Rational Society.* Boston: Beacon Press, 1968.

Martin Heidegger. *The Question Concerning Technology.* New York: Harper and Row, 1977.

Larry A. Hickman. *John Dewey's Pragmatic Technology.* Bloomington: Indiana University Press, 1990 (compiles the most relevant Dewey quotations on technology).

Herbert Marcuse. *One Dimensional Man.* Boston: Beacon Books, 1968.

Karl Marx. *The Marx-Engels Reader* (2nd Edition). Edited by Robert Tucker. New York: W.W. Norton, 1978.

## III. PHILOSOPHY OF SCIENCE AS RELATED TO TECHNOLOGY

Robert Ackermann. *Data, Instruments, Theory.* Princeton: Princeton University Press, 1985.

Peter Galison. *How Experiments End.* Chicago: University of Chicago Press, 1987.

Ian Hacking. *Representing and Intervening.* Cambridge: Cambridge University Press, 1983.

Patrick Heelan. *Space Perception and the Philosophy of Science.* Berkeley: University of California Press, 1983.

Don Ihde. *Instrumental Realism.* Bloomington: Indiana University Press, 1991.

Bruno Latour. *Science in Action.* Cambridge: Harvard University Press, 1987.

## IV. CONTEMPORARY PHILOSOPHY OF TECHNOLOGY

Edward Ballard. *Man and Technology.* Pittsburgh: Duquesne University Press, 1978.

Albert Borgmann. *Technology and the Character of Contemporary Life.* Chicago: University of Chicago Press, 1984.

Barry Cooper. *Action Into Nature: An Essay on the Meaning of Technology.* Notre Dame: University of Notre Dame Press, 1991.

Andrew Feenberg. *Critical Theory of Technology.* Oxford: Oxford University Press, 1991.

James K. Feibleman. *Technology and Reality.* The Hague: Martinus Nijhoff, Publishers, 1982.

Frederik Ferré. *Philosophy of Technology.* New York: Prentice Hall, 1988.

Don Ihde. *Technics and Praxis: A Philosophy of Technology.* Dordrecht: Reidel Publishers, 1979.

———. *Technology and the Lifeworld: From Garden to Earth.* Bloomington: Indiana University Press, 1990.

Nicholas Rescher. *Unpopular Essays on Technological Progress.* Pittsburgh: University of Pittsburgh Press, 1980.

Langdon Winner. *The Whale and the Reactor.* Chicago: University of Chicago Press, 1986.

# INDEX